Bare Bon...

M000273578

A Survey of
Forensic Anthropology

Second Edition

Michael W. Warren, Ph.D., D-ABFA
Nicolette M. Parr, M.S.
Katherine E. Skorpinski, M.A.
Carlos J. Zambrano, M.S.
University of Florida

*Acts of injustice done
Between the setting and the rising sun
In history lie like bones, each one.*

W. H. Auden (1907–1973)

Kendall Hunt
publishing company

All images courtesy of University of Florida, Department of Anthropology and used with their permission

Except the following images:
GPR Readout, courtesy Dr. John Schultz
Gravesite in Bosnia and Morgue in Kosovo, Michael Warren
Forensic Anthropologist, Dr. Laura Fulginiti
Anthropologist, Paul Emanovsky

Cover image © Michael Warren photo.

www.kendallhunt.com
Send all inquiries to:
4050 Westmark Drive
Dubuque, IA 52004-1840

Copyright © 2008, 2011 by Michael W. Warren, Nicolette M. Parr, Katherine E. Skopinski, and Carlos J. Zambrano

ISBN 978-0-7575-8785-6

Printed in the United States of America
10 9 8 7 6 5 4 3 2

TABLE OF CONTENTS

Bare Bones: A Survey of Forensic Anthropology

FOREWORD FOR FIRST EDITION

Two surveyors are marking a boundary line for a proposed industrial park. As they hack through the underbrush along the boundary margins to open a line of sight for their transit, they discover several bones scattered about on the surface of the ground. As they look around, they see more and more bones, until at last they find a skull. The skull seems human! They immediately notify the police, who arrive on the scene. After agreeing that the skull does indeed appear to be human, the police establish a crime scene perimeter and begin an investigation. Whose remains are these? What happened? Was a crime committed? Was a dead body dumped here, or was the person murdered at the scene? These are the types of cases in which a forensic anthropologist may be called to recover and examine the remains. With a thorough investigation by law enforcement and crime scene personnel, consultation between various experts, and a little luck, the identity of the victim and the cause and manner of death can be established, resulting in a successful resolution of the case – both for the legal system and the family and loved ones of the decedent.

This textbook is written to introduce undergraduate students to the exciting world of forensic anthropology, an applied field of biological anthropology. We, the authors, enjoy the privilege of teaching ANT 3522, Skeleton Keys: Introduction to Forensic Anthropology at the University of Florida. The course has become quite popular, drawing students from several colleges around campus and filling the larger lecture rooms of Turlington Hall. Many of the students taking the course are not anthropology majors, so we enjoy introducing them to the world of anthropology and the forensic sciences, and sharing our experiences as forensic practitioners. We expect that few, if any, of the students in our course will pursue a career as a forensic anthropologist. Therefore, we try to shape the syllabus in a way that allows us talk about what we do, demonstrate how it relates to the medicolegal system and society, and show the ways in which anthropology is an important and pertinent pursuit. We do not teach proficiency in forensic techniques in this class – we have other courses that serve that purpose. Instead, we aspire to lead a general discussion about the full range of problems associated with human skeletal identification and trauma analysis, in the hope that our discussions will serve as a model for understanding the broader aspects of biological anthropology and how it helps us to understand the human condition.

The book is divided into three sections that correspond to the three tests given to our students. The first section provides the background for the discipline, frames forensic anthropology within the medicolegal system, and covers the basics of skeletal recovery and osteology. The first chapter frames the field within our medicolegal system and explains how anthropologists articulate with pathologists, odontologists, crime scene investigators and other forensic experts. In the second chapter, we briefly introduce students to the field of anthropology and explain what the role of the forensic anthropologist is within the forensic sciences and academia. Next, we discuss how remains are located, recovered and documented as biological evidence. Chapter four covers the basics of human osteology and acquaints the reader with basic skeletal anatomy and terminology, which sets the foundation for understanding the following chapters.

The second section deals with personal identification from skeletal remains, beginning with the development of a biological profile and ending with the identification of the decedent. Forensic anthropologists create a biological profile from which to begin the search for the identity of the decedent. The first four chapters of this section deal with these in turn: biological sex, ancestry, age-at-death, and stature. Chapter 9 provides information about how investigators can determine personal identity from the skeleton – narrowing down the possibilities until all others are excluded. The final chapter of this section delves into forensic art, a "cold-case" method that generates leads by approximating the physiognomy of the decedent using both science and art.

The third section covers skeletal trauma and other topics related to determining the cause and manner of death, and describes the role of anthropologists in various types of cases ranging from cremated remains to mass disasters. Finally, we conclude our survey with a discussion of the profession of forensic anthropology.

We hope that students find the book enjoyable and interesting. This is our favorite class to teach and we believe that the subject is fascinating and should easily hold your attention.

FOREWORD TO THE SECOND EDITION

This second edition of Bare Bones adds two additional co-authors, Carlos Zambrano and Katie Skorpinski. Both are forensic anthropologists who have been actively involved in forensic casework over the past several years. Both also have taught large section introductory courses in forensic identification. This second edition also corrects some minor errors found by alert students and others over the last two years. We also benefitted from some insightful critiques. Many of the students wanted more photographs of scientists working in the field and laboratory, so we have updated and added several new images in this edition. Another new feature is sample test questions to aid study sessions and encourage further discussion among students, provided by Emilee Amihere.

The authors would like to acknowledge several individuals who have helped with different aspects of putting together the second edition. Foremost, we thank the graduate students of the C.A. Pound Human Identification Laboratory for their dedication to the laboratory's mission. Laurel Freas, Traci Van Deest, Kristina Ballard, Maranda Kles, Caroline Dimmer, and Allysha Winburn are among those who either directly or indirectly contributed to this edition. Merissa Olmer is responsible for almost all of the images in the chapter on human osteology. We also thank the hard work of our highly talented undergraduate volunteers, including Cassie Lucio, Le Hi, and Sebastian Larrea, for their hard work.

Without their professionalism and contribution this book would never have been written.

Section I

Medicolegal Aspects of Death

What interests does society have regarding the death of one of its members? Every culture has a mechanism for dealing with death, including "experts" that are tasked with understanding how and why death occurred, and customs detailing how surviving members of the group should dispose of the body. In our society the investigation of death is the responsibility of medicolegal investigators trained in determining the cause of death and rendering a legal opinion as to the manner of death. The responsibility for this process depends on the nature and context of the death.

Attended deaths

A death is said to have been *attended* when the decedent was under the care of an allopathic or osteopathic physician, and the death was either expected or anticipated based on medical history. An example might be: a patient diagnosed with terminal cancer who is under the care of an oncologist who followed the course of the disease process. In these types of cases, the attending physician is able to rule out other, unexpected causes of death. When the patient expires, the physician signs the death certificate, notifies the funeral home and the body is removed from the home, hospice or hospital.

What happens to a person's body after death? This is not intended as a metaphorical or spiritual question, but rather as a question of practical importance. When a person dies, their body presents several problems for those who remain living. The presence of a body is a constant reminder of the emotional sense of loss experienced by the family and friends of the decedent. In addition, the natural decomposition of the body is unpleasant and poses a health hazard for the living. Fortunately, since ancient times these problems have been solved by burial, cremation and other modes of disposal of remains. In many societies, professional providers – funeral directors or their equivalent – handle almost all of the arrangements pertaining to memorial rituals, preparation of the body, and its burial or cremation.

There is a general progression of events following a death. In most cases, the family of the decedent, often with the help of hospital or hospice employees, will contact a funeral home for removal of the body. If death occurred at home, the family should notify law enforcement officials so they can confirm that the death was a result of natural causes. Many officers will assist the family with contacting the funeral home, and will remain until the body is removed. The body is then transported to the funeral home and, if the family has expressed their desire for a memorial service, is prepared accordingly. If there is to be a memorial service at a later date, the body must be embalmed in order to preserve the tissues. If no memorial service or wake is to take place, the body can be refrigerated until burial or cremation. Some religious groups require burial within 24 hours and prohibit embalming of the body. Apart from burial and cremation – the most common options – there are several other options for disposal of the remains (**Figure 1**).

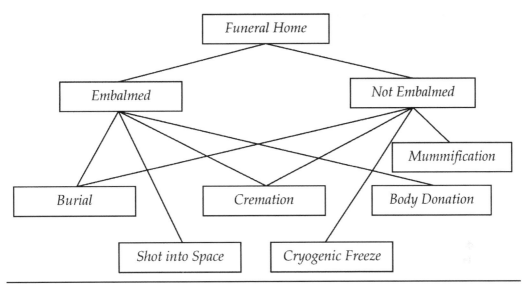

■FIGURE 1: *A schematic showing the possible dispositions of a body in the case of an attended death.*

If the decedent is an organ and tissue donor, the organ-procurement team must harvest the organs and tissues soon after death. Organ donor cards often have a telephone number so that family members and healthcare providers can notify the relevant teams soon after death occurs. If the case falls under the jurisdiction of the Medical Examiner (see below), then the organ-procurement team may harvest the organs and tissues at the Medical Examiner's autopsy suite.

Unattended deaths

A death is **unattended** when it is unexpected, the result of traumatic injury, or when the decedent has no primary physician that might be able to render an opinion as to the cause of death. Examples include car accidents, homicides, or a sudden illness that was not known to have been diagnosed or treated by a physician. Unexpected deaths beg the questions, "How did death occur; and is there a public health or legal issue that should be addressed because of it?" In these cases, responsibility for investigating the death falls to coroners or medical examiners.

THE CORONER SYSTEM

The beginnings of the coroner system dates back to the 11th Century in England, when the coroner served to protect the financial interests of the crown during criminal legal proceedings. Since the 19th Century, coroners have been mostly concerned with determining the cause and manner of death, by conducting an examination of the body and leading inquiries into the circumstances surrounding the death.

Coroners in the United States are usually elected county- or parish-level officers. Although many have medical training, it is not a requirement in many jurisdictions. The authority given to coroners varies by state or region. In some jurisdictions, the coroner has limited law-enforcement powers, such as executing arrest warrants and serving subpoenas requiring testimony. In some cases, the coroner's ability to initiate an investigation, or inquest, exceeds the powers of the medical examiner.

The primary task of the coroner is to determine whether a death falls under his or her jurisdiction, and whether an autopsy should be performed to help determine the cause of death. If an autopsy is to be performed, the coroner will consult with a pathologist – either a hospital-based pathologist or a state Medical Examiner - who will perform the examination and report their findings back to the coroner.

Coroner systems are slowly being replaced with medical examiner's systems, which, as seen below, employ specially trained physicians to determine the identity of the decedent and the cause and manner of death. Coroner systems tend to be preferred in areas where resources are limited; in rural areas where the number of cases requiring investigation is low; and in areas where there are not enough qualified pathologists to institute a medical examiner's system.

MEDICAL EXAMINER'S SYSTEM

Medical examiners are physicians who have received advanced training in a residency program in anatomical and forensic pathology. In Florida, Chief Medical Examiners are board-certified forensic pathologists appointed by the Governor. They provide services within their districts, which are divided among 24 judicial circuits.

Bare Bones: A Survey of Forensic Anthropology

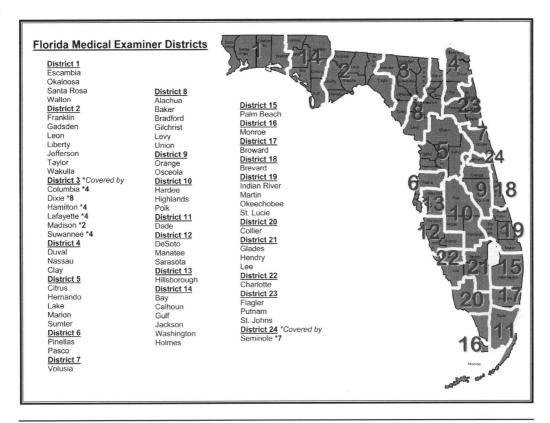

Florida Medical Examiner Districts

District 1
Escambia
Okaloosa
Santa Rosa
Walton
District 2
Franklin
Gadsden
Leon
Liberty
Jefferson
Taylor
Wakulla
District 3 *Covered by*
Columbia *4
Dixie *8
Hamilton *4
Lafayette *4
Madison *2
Suwannee *4
District 4
Duval
Nassau
Clay
District 5
Citrus
Hernando
Lake
Marion
Sumter
District 6
Pinellas
Pasco
District 7
Volusia

District 8
Alachua
Baker
Bradford
Gilchrist
Levy
Union
District 9
Orange
Osceola
District 10
Hardee
Highlands
Polk
District 11
Dade
District 12
DeSoto
Manatee
Sarasota
District 13
Hillsborough
District 14
Bay
Calhoun
Gulf
Jackson
Washington
Holmes

District 15
Palm Beach
District 16
Monroe
District 17
Broward
District 18
Brevard
District 19
Indian River
Martin
Okeechobee
St. Lucie
District 20
Collier
District 21
Glades
Hendry
Lee
District 22
Charlotte
District 23
Flagler
Putnam
St. Johns
District 24 *Covered by*
Seminole *7

■FIGURE 2: *The Florida Medical Examiner Districts. Map is courtesy of the Florida Department of Law Enforcement (http://www.fdle.state.fl.us)*

Medical Examiner's Offices in densely populated areas are often comprised of only a single county, while the more rural areas of the state are composed of multi-county districts. Most offices employ several Associate Medical Examiners, medicolegal death investigators, autopsy technicians and staff to assist in handling their respective case loads. Medical Examiners are given broad investigatory powers and authority by the government, without which they would be unable to perform their jobs. For example, they have the ability to claim jurisdiction over any death deemed to be worthy of forensic investigation; and no permission from any other law enforcement agency or family members is required.

The primary task of the death investigation is to identify the decedent and determine the *cause* and *manner* of death. The cause of death is described as an anatomical diagnosis of the precise mechanism of death. For example, a motor-vehicle accident may result from massive hemorrhage secondary to laceration of the aorta due to rapid deceleration. The manner of death falls into four categories: natural, homicide, suicide, and accidental. In the aforementioned car accident, the manner could be an accident, a suicide (if the accident was intentional), or homicide (if the accident was purposely caused by another driver or through neglect or improper operation of a vehicle – *e.g.* vehicular manslaughter). In some cases, the medical examiner will be able to determine the cause of death, but the manner is recorded as indeterminate.

In addition to performing an autopsy, the medical examiner will review the decedent's medical history, review statements by witnesses, and often examine the scene of death in an effort to glean information that might be helpful in determining the cause of death.

The medicolegal investigation of death involves the efforts of a multidisciplinary team, headed by the Medical Examiner. Members of the team include crime scene investigators, medicolegal death investigators, homicide investigators, state attorneys and prosecutors. This team works to:

- Determine the identity of the decedent

  As will be discussed in Chapter 10, identification of the decedent is important for several reasons. First, it helps to assuage the emotional grief of the decedent's family and friends to know, with certainty, that their loved one is dead. Secondly, it satisfies societal institutions concerned with insurance, probate and other entities that need to record the death of an individual. And finally, if a crime has been committed, knowing the identity of the victim is an extremely helpful tool in prosecuting the offender.

Bare Bones: A Survey of Forensic Anthropology

- Determine cause and manner of death

  The cause and manner of death will, of course, be paramount in determining whether a crime has been committed. The manner of death is also important to the aforementioned social institutions (e.g., the insurance policy may pay benefits if the death was an accident, but not if the death was a suicide).

- Determine the time of death

  This establishes a time line for investigators. When was the victim last seen? In Florida, cases over 75 years old come under the purview of the state archaeologist (a de facto statute of limitations on homicide).

- Identify, collect and preserve evidence

  Forensic pathologists have the authority to retain specimens for further analysis (such as toxicological or genetic studies), for later examination by other experts and, when appropriate, for presentation in court.

- Provide factual information to law enforcement, prosecutors, defense attorneys, families and the public

  Medical Examiners testify in court very frequently. Since the body is the primary evidence of homicide, the jury needs to hear, first hand, the conclusions of the pathologist; and how those conclusions were reached.

- Protect the innocent and prosecute the guilty

  The pathologist, like any scientific expert, does not "work" for either the prosecution or the defense in criminal cases. Instead, they work for the truth. Their testimony may be just as important in exonerating the defendant as in convicting the defendant.

The cornerstone of a death investigation is the ***autopsy*** (from the Greek, literally "to see for oneself"), which involves careful dissection and examination of the organs and tissues of the body. During an autopsy, the pathologist removes and examines all of the organs for signs of trauma or pathology (**Figure 3**).

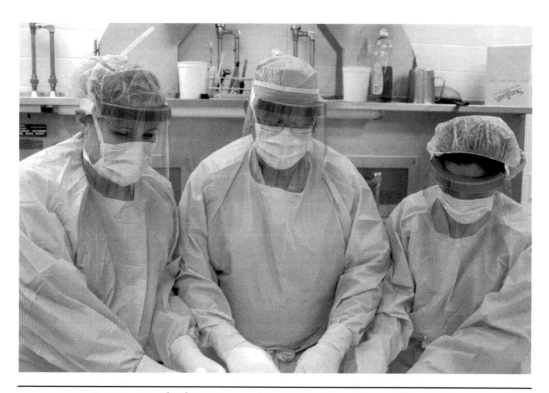

■FIGURE 3: *Charlotte Stevens, Dr. William Hamilton, and Sarah Thomas perform an autopsy to determine the cause and manner of death.*

Medical Examiners are expert in describing and recording all pertinent information that might be called into question during a homicide trial. They not only note relevant findings, but also anticipate which negative findings may be important. Small samples of tissue are always retained for further analyses, including histological examination, toxicology and DNA testing.

Families of decedents who died under the care of an attending physician may also request an autopsy. An autopsy may confirm or refute the working diagnosis on which treatment was based, determine the severity of the illness, and address the possibility of malpractice; or reveal genetic or familial diseases of interest to related family members. Fetal autopsies are routine for most cases of spontaneous abortion. The autopsy may reveal the cause of the miscarriage and alert the mother to any genetic anomalies that should be considered before another pregnancy. Some diseases can only be confirmed after autopsy by visual and histological inspection of the tissues (*e.g.*, Creutzfeldt-Jakob

Bare Bones: A Survey of Forensic Anthropology

Disease, a type of spongiform encephalopathy). These autopsies are usually performed in the hospital by clinical pathologists, many of whom do not have training in forensic pathology.

How do forensic anthropologists fit into this investigation? They are just one of several types of experts with whom the Medical Examiner can consult under special circumstances. The Venn diagram in Figure 4 shows the relationship between several disciplines, any representative of which may be asked to assist in the recovery, identification and trauma analysis of skeletonized remains. The diagram places the Medical Examiner in the center, since he or she has the legal mandate to determine the cause and manner of death. However, other disciplines may be called upon to assist. An anthropologist may be asked to assist in recovering buried or scattered remains, and to help identify and assess skeletal trauma in cases involving skeletonized, decomposed, burned, or fragmented remains. A forensic odontologist is often consulted to help identify the decedent based on a radiographic match between a specimen and existing dental records. Law enforcement officers, crime scene technicians, archaeologists, or medicolegal death investigators collect material evidence. Perhaps bone samples will be sent to a genetics laboratory for mitochondrial DNA sequencing and matching.

A whole slew of other forensic experts are available to the pathologist, including entomologists, radiologists, botanists, and others with specialized training. All of these experts work together in an effort to answer all questions and resolve the case. The next chapter, as well as the remainder of this book, is devoted specifically to the forensic anthropologist's role within this multi-disciplinary team.

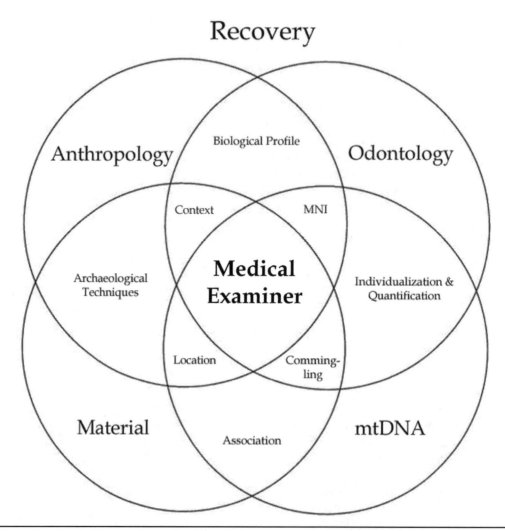

FIGURE 4: *A Venn diagram showing how various disciplines articulate during the investigation of death in a case involving skeletonized remains (with thanks to Drs. Tom Holland, Bob Mann and John Byrd; Joint POW/MIA Accounting Command, Central Identification Laboratory).*

Sample Test Questions

1. **A decedent has a fatal knife wound to the chest. Which are four possibilities as to the *manner* of her death?**

    a. Homicide, accidental, natural, indeterminate

    b. Homicide, judiciary, fratricide, accidental

    c. Homicide, suicide, accidental, indeterminate

    d. Suicide, natural, indeterminate, accidental

2. **Coroners differ from Medical Examiners in that:**

    a. Coroners do not have the legal mandate to determine cause and manner of death

    b. Coroners may not be pathologists, or even physicians

    c. Coroners do not have the authority to conduct death investigations

    d. Coroners also serve as funeral directors for all legal cases in their jurisdiction

3. **Medicolegal death investigators are:**

    a. Law Enforcement Officers assigned to the Homicide Division

    b. Usually civilians who work for the Medical Examiner's Office

    c. Crime Scene Investigators (CSI) specializing in death scenes

    d. Pathologists who specialize in determining the cause of death from clues at the scene

4. In Florida, the retention of biological specimens by Medical Examiners is permitted:

   a. With written consent of the family

   b. With consent of the funeral director

   c. By state law, or statute

   d. Only under court order

5. Medical Examiners and coroners perform autopsies on behalf of the state in order to determine the cause and manner of death. What is/are the reason(s) why the public needs to know the cause and manner in which a person died?

   a. Public safety concerns (*e.g.,* faulty automobiles, disease epidemics, homicide rates)

   b. Family concerns (*e.g.,* discovery of hereditary or familial diseases, emotional closure)

   c. Legal concerns (*e.g.,* prosecution of the guilty, insurance claims, liability issues)

   d. All of the above

What Is Forensic Anthropology?

WHAT IS ANTHROPOLOGY?

Anthropology is the study of humans and the human condition (from the Greek: *Anthropos, meaning* man + -ology; to speak of, or study of). Anthropology is a science, and therefore uses as its basic tool the **scientific method**. As scientists, we search for testable explanations for phenomena that are observed in nature, and seek to find correlations and causations that explain observed phenomena. A theory is an explanatory statement supported by a body of facts. The explanatory statement is predictive, that is, it helps us to know what might happen under similar circumstances. Scientists tell stories that can be disproved. The strengthening or weakening of a theory is the way in which science progresses. Anthropology, as a science, does not claim to seek ultimate truths; that's the job of other departments on campus – such as religion, literature and the arts! But, as with all sciences, anthropology can provide us with a world view that makes us better-informed, more tolerant, fulfilled and happy people.

Anthropology is the most holistic of the social sciences, integrating every facet of life – religion, ritual, economics, political systems, language and others – in order to better understand humankind's place in the universe. All of these components of human society are transmitted by **culture**. E.B. Tylor, an early anthropologist, defined culture as, "that *complex whole* which includes knowledge, belief, art, morals, law, custom, and any other capabilities and habits acquired by man as a member of society" (1871). Culture is acquired through learning and, inexorably bound to culture is human biology, which makes culture possible by virtue of humankind's special brand of intellect, capacity for abstract thought, ability to transmit knowledge through language and literature, and myriad other adaptations unique to the human species.

Anthropology is divided into four broad categories, or subdisciplines: **Socio-cultural anthropology, linguistic anthropology, archaeology** and **biological anthropology**. Each subfield considers different aspects of the human condition, as well as the interplay between these various aspects. Many anthropologists work beyond the boundaries of each subdiscipline, integrating biology and culture, or culture and language, for example. Others work at the margins of anthropology, merging anthropological theory with outside interests, such as geography, chemistry, religion or political science.

Socio-Cultural Anthropology

Socio-cultural anthropology is a field-based science that utilizes direct observation of cultures using a method called **ethnography** – the descriptive study of a particular society. Anthropologists often live among a societal group for a year or more (and sometimes, an entire career), collecting information and data directly from members of the group. The anthropologist collects data through interviews or a series of detailed, repetitive observations and then analyzes those data using a variety of both qualitative and quantitative methods.

Linguistic Anthropology

Anthropological linguistics – or, to use the newer, more accurate term, linguistic anthropology – is the systematic study of human language and communication, with particular focus on the interaction of language and culture. Early anthropological linguists were primarily interested in recording and preserving rare and endangered languages, using ethnographic methods within a linguistic theoretical framework. Linguistic anthropologists are interested in using anthropological paradigms to better understand the historical (and proto-historical) development of languages and language groups, the sound and structure of language, and the influence of language on thought and society.

Archaeology

Archaeology (literally, the *study of ancient things*) is the study of past human life and culture through the examination of material culture – the objects left behind by groups that (in most cases) left no written record of their societies. Archaeologists recover, record, analyze, and classify archaeological material with the goal of describing and interpreting the patterns of human behaviors that led to each artifact's creation (*The Concise Oxford Dictionary of Archaeology PAGE 28*). This, in turn, leads to a better understanding of the reasons for this behavior (**Figure 1**).

■**FIGURE 1:** *Archaeologists during excavations at Kingsley Plantation, Florida. Photograph courtesy of Dr. James Davidson.*

What Is Forensic Anthropology?

Archaeologists do not, however, always examine the artifacts of past civilizations. Historical archaeologists reconcile the written history of a people with the archaeological evidence to better understand the living conditions and lives of relatively recent societal groups. Some archaeologists examine the discarded material culture of extant populations, such as comparing findings from a town landfill with information gleaned from questionnaires about the town's current and historical consumption of a particular type of goods (*e.g.* alcoholic beverages).

Biological, or Physical Anthropology

Biological anthropology is the study of the physical and chemical characteristics of humans and their closely related ancestors. The various subfields of biological anthropology seek answers to questions about humankind's evolutionary history, unique biological adaptations to the environment, the range of physical variation between and within populations, and the mechanisms for such diversity – in short, we are interested in those physical aspects of humans that make us unique among the animals of the world. How has natural history shaped the human animal? What can we learn about the interplay between biology, ecology, and culture? How can the similarities and differences between populations tell us about the evolutionary history of groups (*e.g.* population bottlenecks, large-scale migrations)? Many biological anthropologists place special emphasis on bones and teeth, since these tissues are preserved and are the only surviving biological evidence of past populations **(Figure 2)**.

■FIGURE 2: *Anthropologist Casey Self performs an experiment, using in vitro strain gauge analysis, to better understand the properties of the human mandible.*

A Brief History of Biological Anthropology

Interest in human variation began to emerge in the early 15th Century, during what has been coined The Age of Discovery. During this period, explorers from Europe (intent on increasing trade and finding new riches) began to visit isolated areas of the world that were previously unknown to them. Armed with better ships and navigational tools, they found themselves face to face with people who appeared quite different from themselves – people who talked differently, had different customs and clothing, and in general were surprisingly different in many ways. These differences were difficult to explain, and ultimately led to attempts to categorize and classify populations into groups based on a wide variety of random physical and social characteristics.

The scientific study of the physical characteristics of humans developed during the 19th Century, prior to Wallace and Darwin formulating the theory of natural selection. The field was primarily devoted to the study of physical evidence, such as fossils and measurements of living subjects, and so was called "Physical Anthropology." After the rise of Darwinian theory and the synthesis of natural selection with population genetics, the term "Biological Anthropology" gained favor, and is still used by most anthropologists today.

The early history of biological anthropology can, unfortunately, be associated with antiquated and much-maligned theories of racial typologies. In an effort to explain differences between groups of people, theories emerged that sought to categorize groups of individuals into physical types. These types were believed to be associated with unrelated attributes, such as intellect, criminal behavior, and personality traits. This inevitably led to rankings of groups based on a perceived hierarchy. These rankings were, of course, used to justify the subjugation and extermination of the least desirable types – leading to the eugenics movement, which aimed at creating a perfect race and eliminating inferior lineages.

Biological anthropology today is less interested in the differences between populations than in understanding the full extent of human variability, and how that variability is expressed and directed towards better adaptability across a wide range of environments.

HISTORY OF FORENSIC ANTHROPOLOGY

The relatively brief history of forensic anthropology has been divided previously into three periods by Douglas Ubelaker: The Formative Period, the Consolidation Period, and the Modern Period (Ubelaker and Scammell, 1992). These divisions have since been used by others and seem to have gained traction as a valid and useful way to punctuate the timeline of the many landmark events in the discipline (*e.g.*, Byers, 2005). A Fourth Era has been recently added by Paul Sledzik and colleagues, who see significant change in the new millennium that warrants a nod in discussions about the historical eras of forensic anthropology.

The Formative Period (1849–1938)

The story of forensic anthropology begins in 1849, with the famous Parkman Case. Dr. George Parkman was a prominent Boston physician who later donated the land where the Harvard Medical School sits today. Parkman loaned a sum of money to Dr. John Webster, a chemistry professor at the university. Rather than repay the debt, Dr. Webster murdered Parkman, dismembered his body, burned the head in the building's furnace, and hid the remainder of the body parts in his anatomy laboratory. Two distinguished anatomists, Drs. Oliver Wendell Holmes and Jeffries Wyman, were able to re-assemble the remains and determine that the body was that of a man matching the stature, race and age of Parkman. They also matched the dentures found in the furnace to a set of molds taken by Dr. Parkman's dentist when the dentures were manufactured. The identification of the remains was instrumental in convicting Dr. Webster of the murder. This is the first such case in literature in which examination of skeletal and burned remains was used to identify the victim in a murder trial. This case marks the beginning of what has been termed the Formative Period in the history of forensic anthropology (Byers, 2005). The pracice gained in popularity when, ten years later, Darwin published *On the Origin of Species*, which piqued interest in the evolutionary history of humans, spurring greater interest in research involving the examination of skeletal remains (Darwin, 1859).

In the late 19th Century, Thomas Dwight became the Parkman Professor at Harvard University (named for the very Parkman that was murdered), succeeding Oliver Wendell Holmes at that position. When Thomas Dwight came to Harvard, he had no doubt heard about the case. In 1894, he gave the famous *Shattuck Lecture*, the first truly forensic lecture about the analysis of human skeletal remains in a legal context. Dwight is generally heralded to be the "Father of Forensic Anthropology".

By the first two decades of the 20th Century, anatomists were beginning to develop large series of skeletal collections in order to accomplish their research objectives. Two primary collections were begun during this period, which are still used by researchers today. T. Wingate Todd, a physician, established what is now known now as the Hamann-Todd collection, curated at the Cleveland Museum of Natural History. While teaching anatomy at Western

Reserve University, Todd built a collection of over 3,000 human skeletons. The specimens constitute a "known collection", meaning that each cadaver is accompanied by supporting documentation as to demographic information including height, age at death, and other important data. Carl Hamann, a former Dean of the Western Reserve University School of Medicine, assisted Todd in enlarging the collection and was, therefore, acknowledged in the naming of the collection.

A second, equally important collection is the Terry Collection, housed at the Smithsonian Institution in Washington, D.C. This collection was established by Dr. Robert Terry, and his successor, Mildred Trotter, while at Washington University in St. Louis, Missouri, between 1914 and 1965. More than 1,600 skeletons are included in the collection. The collection is currently under the watchful eye of Dr. David Hunt, who coordinates visiting scholars and provides access to the more than 33,000 specimens housed in the Physical Anthropology Division of the Smithsonian Institution (Hunt, 2008).

These two collections have served as a normative sample for a multitude of research projects related to forensic identification, and they have been an invaluable part of the recent history of forensic anthropology. There are, however, some inherent biases introduced by such collections. For instance, most of the skeletons are procured from cadaver rooms, where they were donated for medical research. Most of the individuals were of low socio-economic status, and so suffered from poor nutrition and health. Therefore, the data collected from the skeletons may not represent a perfect reference sample for the general population. Also, these collections are aging. Populations experience a phenomenon known as secular change – in size and shape – over time. These changes are not genetic, but are rather a result of different (and usually better) nutrition, healthcare, and sanitation. For example, several studies have shown that we are getting progressively taller over time, and that successive generations of immigrants are significantly taller than their foreign-born parents and grandparents (*e.g.*, Garn, 1987; Meadows and Jantz, 1995).

Consolidation period (1939–1971)

The Consolidation Period is so-called because it takes the formative work cited above, and consolidates the discipline through continuing research and practice. The period can be said to begin with Wilton Krogman's *FBI Law*

Bare Bones: A Survey of Forensic Anthropology

Enforcement Bulletin report (Krogman, 1939). In 1962, Krogman would author *The Human Skeleton in Forensic Medicine*, a landmark reference text that, to this day, resides on every forensic anthropologist's bookshelf (Krogman, 1962).

The beginning of World War II in 1939, and America's involvement starting in 1941, signaled a new era in forensic identification. The thousands of casualties who were either lost or left on the field of battle led to a massive effort to locate, repatriate and identify each and every soldier, sailor and Marine. At the close of the war, the U.S. Army established the Central Identification Laboratory under the direction of Charles Snow, and later, Mildred Trotter. This laboratory, which still operates today under the Joint POW/MIA Accounting Command (JPAC) in Hawaii, has the sole mission of identifying servicemen lost in prior wars. The JPAC/Central Identification Laboratory stands today as the largest such laboratory in the world, employing over 30 professional forensic anthropologists. Servicemen lost during World War II, the Korean War, and the Vietnam War are still being repatriated, identified and returned to their loved ones – some, over 60 years after their death.

The Central Identification Laboratory (CIL) initially faced the tremendous task of identifying skeletal remains with little scientific data to aid in their efforts. However, the war dead not only presented scientific problems, but presented scientific opportunity as well. In order to determine the age-at-death, stature and race of the individual's remains under their care, CIL scientists were required to collect reference data that would permit them to accomplish their goals. Several seminal studies originated from these early efforts to identify our servicemen. One of the most important was Tom McKern's and T. Dale Stewart's "Quartermaster's Report," titled *Skeletal age changes in young American males*. This report established important aging techniques based on data collected from the identified remains of servicemen from the Korean War (McKern and Stewart, 1957).

The early 1960s provided a new tool for metric analysis of the skeleton. The computer enabled researchers to evaluate large volumes of data and apply very sophisticated techniques to complex problems. In 1962, Giles and Elliot published their first paper using a statistical technique called discriminant function analysis (see Chapter 5). Almost all forensic anthropologists practicing today use this statistical method to aid in identifying the sex and ancestry of unknown skeletal remains.

Modern period (1972–1999)

The modern period began with the establishment of the Physical Anthropology Section of the American Academy of Forensic Sciences (AAFS). The birth of the Physical Anthropology section signaled the recognition that forensic anthropology was ready to take its place among the forensic sciences as an acknowledged contributor to medicolegal investigations. As discussed in the final chapter, the Physical Anthropology Section provides a critical venue for the dissemination of research, exchange of ideas, discussion of case studies, and a general forum for the profession. The formation of the section was followed, in 1977, by the institution of a certifying body – the American Board of Forensic Anthropology (ABFA) – that serves to set a professional standard for practitioners (the ABFA is discussed more fully in the final chapter, as well).

In 1979, Smithsonian Institution anthropologist T. Dale Stewart published a landmark textbook, *Essentials of Forensic Anthropology*, that provided a synthesis of the current methods and techniques employed by practicing forensic anthropologists. The next year, the University of Tennessee opened the first decomposition research facility in the country. Established under Dr. William Bass, the "facility" was to pave the way for taphonomic studies in the decomposition of human remains.

Up until the 1970s, pioneers of forensic anthropology had been trained primarily as skeletal biologists, bioarchaeologists, anatomists, or as scientists in some other, related discipline. These scholars built and defined forensic anthropology as we know it today. During the modern period, they began to institute formal academic training directly related to research in forensic identification. Hallmark programs sprang up at the University of Kansas, University of Tennessee, University of Arizona, University of Florida, and other institutions. These early programs trained many of the forensic anthropologists actively engaged in casework today.

The University of Tennessee's Dr. William Bass and colleagues are single-handedly responsible for training a major portion of the *c.* 75 board-certified forensic anthropologists in the United States. Another notable accomplishment by the faculty of the University of Tennessee was the creation, in 1986, of the Forensic Data Bank. Drs. Richard Jantz and Stephen Ousley later used these data, collected from identified cases submitted by practitioners around the country, to develop a computer software program called FORDISC, that enables investigators to perform metric analyses based on a contemporary reference population (Jantz and Ousley, 1993).

During the same year that the Forensic Data Bank was being developed, Dr. Clyde Snow began to establish and train human rights teams through the American Association for the Advancement of Science (AAAS), starting with the Argentine Forensic Anthropology Team (EAAF). Another noted forensic anthropologist, Dr. William Maples, was busy that year, founding the C.A. Pound Human Identification Laboratory at the University of Florida, the first privately funded laboratory devoted exclusively to the practice of forensic anthropology **(Figure 3)**.

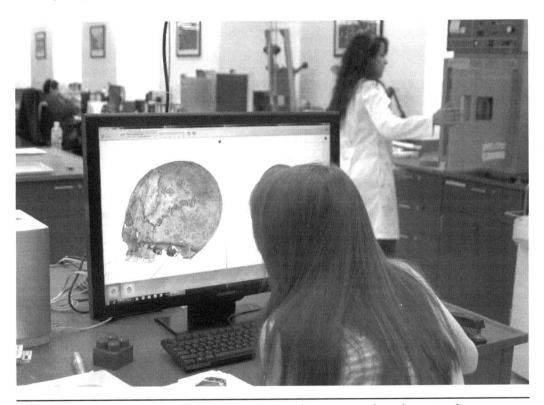

■FIGURE 3: *Caroline Dimmer recording biological evidence using a three-dimensional scanner at the C.A. Pound Human Identification Laboratory, University of Florida.*

What Is Forensic Anthropology?
25

The Fourth Era (2000-present)

The new millennium has brought about several new developments in the field of forensic anthropology. The past few years have seen broadened research goals and new techniques, moved the application of anthropology into the international arena, and created the need for burgeoning education programs – particularly at the graduate level – to prepare new generations of anthropologists for forensic practice. In 2000, the *Ellis Kerley Foundation* was established as the first funding source solely for research in forensic anthropology and closely related disciplines. The Kerley Award has become a much-coveted and prestigious honor among the student members of the AAFS.

New rules of testimony and evidence have prompted a move towards accreditation and certification of individuals, laboratories and educational programs. Certification boards, like the ABFA, are being called upon to become members of umbrella accreditation boards that provide additional oversight and serve to validate the credentials tendered to their diplomates. In 2003, JPAC/CIL became the first skeletal identification laboratory to obtain ASCLD-LAB (American Society of Crime Laboratory Directors-Laboratory Accreditation Board) accreditation. Many forensic science programs are seeking accreditation from the Forensic Education Program Accreditation Commission (FEPAC). Are forensic anthropology programs next?

The future of anthropology has been the subject of a great deal of discussion among colleagues within the discipline, with professional symposia organized to permit debate about the direction of educational programs, accreditation, and other issues of importance to the profession. We will discuss some of these issues further in the final chapter of this book.

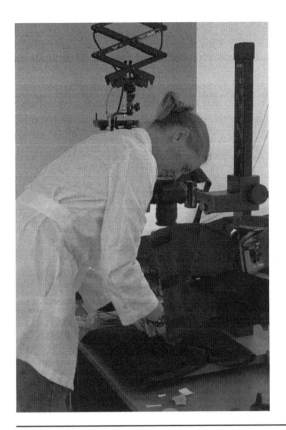

■FIGURE 4. *Traci Van Deest taking forensic photographs to document evidence and use as courtroom exhibits.*

References

Byers SN (2005) *Introduction to Forensic Anthropology: A Textbook, 2nd Edition.* Boston: Allyn & Bacon.

Darwin CR (1859) *On the Origin of Species by Means of Natural Selection, or the Preservation of Favoured Races in the Struggle for Life.* London: J. Murray.

Garn SM (1987) The secular trend in size and maturational timing and its implications for nutritional assessment. *Journal of Nutrition* 117(5):817–823.

Hunt D (2008) History and Collections of the Division of Physical Anthropology, National Museum of Natural History, Smithsonian Institution. In Warren MW, Walsh-Haney HA, Freas LE (eds.) *The Forensic Anthropology Laboratory*. Boca Raton, Fla.: Taylor & Francis.

Jantz RL and Ousley S (1993) *FORDISC 1.0*. Department of Anthropology, University of Tennessee.

Krogman WM (1939) A guide to the identification of human skeletal material. *FBI Law Enforcement Bulletin*, Volume 8, No. 8.

Krogman WM (1962) *The Human Skeleton in Forensic Medicine*. Springfield, Ill.: Charles C. Thomas.

McKern TW, Stewart TD (1957) Skeletal age changes in young American males. *Technical Report No.* EP-45; QMT Research and Development Center, Environmental Protection Research Division, Natick, Massachusetts. Also known as the "Quartermaster's Report."

Meadows L, Jantz RL (1995) Allometric secular change in the long bones from the 1800s to the present. *Journal of Forensic Sciences* 40(5):762–767.

Sledzik PS, Fenton TW, Warren MW, Byrd JE, Crowder C, Drawdy SM, Dirkmaat DC, Finnegan M, Fulginiti LC, Galloway A, Hartnett K, Holland TD, Marks MK, Ousley SD, Rogers T, Sauer NJ, Symes SA, Tidball-Binz M, Ubelaker D (2007) The Fourth Era of Forensic Anthropology. *Proceedings of the American Academy of Forensic Sciences* 13:350–353.

The Concise Oxford Dictionary of Archaeology. Oxford University Press, 2002.

Tylor EB (1871) *Primitive Culture*. London: J. Murray.

Ubelaker DH, Scammell H (1992) *Bones, a Forensic Detective's Casebook*. New York: M. Evans and Company.

Sample Test Questions

1. Which of the following is NOT one of the sub-disciplines of anthropology?

    a. Archaeology

    b. Ethnography

    c. Biological anthropology

    d. Socio-cultural anthropology

2. Ubelaker divided the history of forensic anthropology into all the following periods except:

    a. Modern period

    b. Formative period

    c. Consolidation period

    d. Millennial period

3. Who gave the Shattuck Lecture in 1894?

    a. T. Dale Stewart

    b. Dr. George Parkman

    c. Thomas Dwight

    d. Dr. Oliver Wendell Holmes

4. Who founded the C.A. Pound Human Identification Laboratory at the University of Florida?

    a. Dr. William Maples

    b. Dr. Clyde Snow

    c. Dr. Richard Jantz

    d. Dr. Joseph Hefner

Recovery, Documentation, and Preparation of Human Remains

The work of the forensic anthropologist begins when skeletonized, decomposed, or otherwise unidentifiable human remains are found. Often, remains are found after exhaustive investigative work by law-enforcement agencies. These investigations may have identified possible victims, perpetrators, and witnesses, which then led to the location of the victim's remains. Just as often, hunters, surveyors, berry pickers, or others find remains accidentally.

Skeletal remains are found in two general forensic contexts, enclosed and open. **Enclosed scenes** include houses, railroad cars, dumpsters, or any scene in which the body is confined to a known area. **Open scenes** include scenes in which bodies have been left on the surface of the ground—often scattered or buried in graves. Since clandestine graves are often rapidly dug and without proper tools, it is not uncommon to find a combination of a shallow grave with some surface scatter of skeletal remains due to erosion or animal scavenging. Water recoveries are another type of open scene that requires specialized knowledge and techniques.

This chapter will briefly cover the general principles of search and recovery, forensic archaeology, and the documentation and processing of human remains. For a more detailed description of the methods and techniques of forensic archaeology, readers may consult Dupras and colleagues' excellent book on location and recovery of skeletal remains (Dupras et al. 2006).

THE SCENE

The location where a dead human body is found may provide some of the most important evidence related to the crime. Human remains are crucial biological evidence and represent the *corpus delicti*—the primary evidence that a crime has been committed. The scene represents a place in time when the perpetrator and victim were together. Methods used to locate and recover the remains are designed to preserve and record the crime scene so that the least amount of evidence is lost. The goals of the search and recovery are as follows:

- Find the location of the remains using the least invasive techniques possible.

- Secure the scene so that the evidence remains untainted and in as pristine condition as when found.

- Record where the remains were found.

- Record the relationship between various elements of the scene including bones, clothing, weapons, or other evidence.

- Discover the method of body disposal (i.e., was the victim killed at the scene, or was the body dumped?).

- Properly remove all of the biological evidence from the scene.

- Properly map the site so that the crime scene can be reconstructed at a later date.

THE SEARCH

Prior to conducting a search for human remains, there must be reason to believe that a body is in the area. Perhaps a suspect or witness has confessed to the crime and has led law-enforcement officers to the scene. Witnesses may have seen suspicious activities that they have connected to a crime, or maybe the victim was last seen in a wooded area near his or her home. In cases of surface depositions where skeletal remains have been found, a search is still required to make sure that the anthropologists and crime-scene technicians have recovered *all* of the remains present.

The first task is to formulate a search plan that addresses all of the challenges and considerations for the particular crime scene. Investigators may begin by consulting topographic maps or aerial photographs of the area to gain better understanding of the terrain and identify obvious obstacles. Aerial photographs also provide an overview of roads and paths that might have been used as ingress and egress by the perpetrator. Is access to the scene limited? Might the perpetrator have had prior knowledge of this secluded area?

Large sites may begin with a minimum number of experienced searchers who can reconnoiter the area, establish the search parameters, determine the types of additional equipment and resources needed, and decide which type of search pattern would most likely yield results.

Search methods may be labeled as either noninvasive or invasive. **Noninvasive search methods** include satellite or aerial imaging, remote geophysical imaging, detection, line, and grid searches, and cadaver dogs. These methods, if properly employed, are not destructive to the scene—although it may be argued that the mere presence of searchers alters the environment to some extent. **Invasive search methods** are those methods that, to some degree, irreversibly alter the crime scene and have the potential to damage evidence. These methods include soil probing, test pits, and trenches. Common sense dictates that it is always best to begin with noninvasive methods first and then proceed with invasive techniques if the noninvasive methods are unsuccessful.

Noninvasive Search Methods

The least invasive search methods are those in which no one is actually at the scene! Anyone familiar with the Web site Google Earth (earth.google.com) will know that relatively high-resolution satellite imaging is both available and easily accessible. Investigators may consult publicly available computer programs such as Google Earth or Google Maps, or they may get aerial photographs from government sources, such as a county tax assessor's office that uses small aircraft flybys to take photographs to help assess property values and resolve zoning issues. Many law-enforcement agencies own or operate small planes or helicopters that can be enlisted to provide search support. In special circumstances, it may be possible to gain access to military satellite images. For example, the U.S. military provided information about the location of several mass gravesites in Bosnia following the Balkan war in the mid-1990s.

Once the parameters of the search area are set, investigators can begin to walk the site. One of three types of search patterns is usually used. A **line search** involves from a few to several people slowly advancing forward in a line. If everyone stays at arm's length and proceeds forward slowly and in unison, a large amount of ground can be thoroughly and effectively searched. Once the line has progressed across the area to be searched, the line then "flops," and the search proceeds back across the site. Small marker flags are often used to make sure that no area is missed. A **grid search** is an extension of a line search, but instead of a single pass, the group then searches the same area— but this time moving in a line perpendicular to the first pass (e.g., if the first line search moved north to south and returned south to north, the second pass would move east to west and back). See **figure 1**. A **circular search** may be used for finding and mapping skeletal remains that are found underwater. First, a baseline is stretched along a cardinal direction. The baseline is a rope with knots tied every five meters. A secondary line, with knots every meter, can be clipped to the baseline at five-meter intervals. The searcher then starts at the first meter knot and makes a circular sweep from baseline to baseline. Then, the searcher proceeds to the two-meter knot and performs another sweep. The diver can map the location of any object by the number of meters along the baseline, plus the number of meters out from the swing line and its cardinal location relative to the baseline (e.g., fifteen meters north of the datum, three meters out from the west side of the baseline).

Bare Bones: A Survey of Forensic Anthropology

■FIGURE 1: *Anthropologists conducting a shoulder-to-shoulder line search during a field school at Mercyhurst College, under the direction of Dr. Dennis Dirkmaat.*

Any of these search methods are likely to find scattered human remains, provided that each member of the team is familiar with the appearance of human bones and is thorough and meticulous while searching for remains. The remains are sometimes readily apparent and confined to a small scene.

■FIGURE 2: *A surface deposition that covers a relatively small area. The remains have been minimally scavenged*

Once remains are found on the surface, investigators flag each element of evidence **in situ**. As the search progresses and more and more bones are found, the flags make it possible for investigators to better see there is a pattern to the dispersal of the skeleton. Each bone and piece of evidence is carefully photographed and mapped prior to being removed from the scene. Once a bone is moved, it can never be placed exactly as it was! Once each element is carefully documented in situ, then it is tagged, assigned a number, and placed into evidence bags. The flag that marked the location of the bone is often left in place, in case personnel wish to return to the scene for further investigative work.

Bare Bones: A Survey of Forensic Anthropology

■FIGURE 3: *The early stages of recovery of a surface deposition*

What if the remains have been buried? Clandestine, criminal graves require additional search and recovery techniques. The location, mapping, and recovery of human remains involve the subfield of forensic archaeology. **Forensic archaeology** is the adaptation of standard archaeological theory and procedures to crime scenes involving buried bodies or skeletal remains.

Locating Clandestine Burials

Evidence for burial may be geological, environmental, artifactual, or a combination of these three categories. The visual signs of a grave include a surface depression or mound. Graves in which the body has yet to decompose may have an excess of soil, since the body is displacing a significant volume of backfill. The backfill is also less compacted than the surrounding soil. Graves in which the body has had sufficient time to decompose may have collapsed into a shallow depression. It is difficult for one to dig a hole and then afterwards return all of the displaced soil. Over time, the soil in the grave shaft will become compacted again, and the body—now reduced to a skeleton—will no longer displace the soil within the grave.

Graves also produce changes in soil color. Soils are stratified. In a large hole, soils from different stratigraphic layers are mixed in the backfill. When the backfill is returned to the hole, the mixed soils will be a different color than the undug, virgin soils nearby. Also, the lipids and volatile exuded from a decomposing body invariably turn surrounding soils a darker color.

Shallow graves also affect vegetation. The vegetation at the gravesite is dug up or otherwise damaged. The digger may have been forced to cut through roots or remove rocks and stones from the grave shaft. Once the body is placed into the grave, it may take some time before the vegetation returns. On the other hand, a decomposing body produces large amounts of nitrogen and other plant nutrients. It is not uncommon to see a disturbed area of soil, surrounded by bright green vegetation that has grown significantly more than the surrounding plant life.

The dead body is also a source of nutrition for animal life in the area. Dogs, coyotes, and other carnivores or scavengers often breach shallow graves. This animal activity may enable investigators to see bones, clothing, or other evidence on the surface. If the anthropologist pays careful attention to the distribution pattern of these bones and artifacts, it may be possible to locate the gravesite.

Even if a suspected grave has been found, it is best to confirm the finding and delineate the grave parameters with noninvasive techniques prior to probing or digging. One noninvasive method is **ground-penetrating radar**, or GPR. This method uses a radio transponder and receiver to send radio waves into the ground and record returning echoes from anomalies beneath the surface **(Figure 4:** GPR readout).

Bare Bones: A Survey of Forensic Anthropology

■FIGURE 4: *This GPR printout records two anomalies. Both of these signals represent experimental pig burials at 0.5 meters deep. Photograph courtesy of Dr. John Schultz.*

Ground-penetrating radar has been in use by soil scientists, geologists, and others for many years; however, forensic scientists are only recently performing research on remote sensing for clandestine graves. Schultz has shown that GPR is quite effective in locating and delineating shallow graves of varying depths (Schultz 2008a, 2008b).

Ground-penetrating radar requires special equipment and personnel but is well worth the effort, since the next steps are likely to involve destructive processes. GPR is more effective in sandy soils than rocky soils, since the investigator will sense fewer subsurface anomalies.

Another remote-sensing tool is **magnetometry**, which locates changes in the magnetic field due to metallic objects. This method is not intended to locate bodies but instead is used to find guns, knives, belt buckles, or other metallic objects that might have been left in the grave.

Some investigators have had great success with **infrared photography** and imaging. Infrared cameras use the infrared spectrum of light to detect heat. This method has been widely used to detect living people in manhunts and disasters and is particularly effective at night, when the air is cooler in relation to the body. Infrared can also be used to detect decomposing bodies, which produce heat during the decomposition process. The backfill soils within a grave are less compact than the surrounding soils, causing less heat retention. Clandestine graves will show up as a "cool spot" when viewed through an infrared camera.

A final remote-sensing method involves **cadaver dogs**. These dogs are specially trained to detect the scent of the byproducts of decomposition. Once the dog detects a source, it "alerts" its handler that it has "hit" on a possible cadaver. Most practicing forensic anthropologists have achieved mixed results using cadaver dogs. Some cadaver dogs have an outstanding record; others have not fared as well.

FORENSIC ARCHAEOLOGY

As defined above, **forensic archaeology** is the application and adaptation of traditional archeological methods to meet the needs and requirements of crime scenes. The theories underlying the work of forensic archaeologists are the same as those used by traditional archaeologists. However, forensic crime scenes place several constraints on the forensic archaeologist that call for an adaptive approach. Special considerations faced by forensic archaeologists include maintaining security of the crime scene for up to several days, the urgency of recovery and identification of the body in order to resolve the case, and dependence on unreliable sources of information as to site location. These factors generally mean the forensic archaeologist is afforded a much smaller time to locate, map, and recover the remains than traditional archaeologists. Forensic archaeologists must be aggressive in their approach, sometimes using heavy machinery and other protocols that belie the training they received during field schools at established archaeological sites.

Bare Bones: A Survey of Forensic Anthropology

Mapping the site

Once remains are found, the next step is to carefully document the location of the site. The methods and techniques used in mapping depend on the type of scene, the amount of time available for recovery, the resources of the recovery team, and the amount of documentation required by the investigators.

The investigators need to know two things: (1) the location of the remains in space—that is, where were the remains found in the world? and (2) the relationship of the remains to each other and to other landmarks or evidence found at the scene. The location in space may be recorded by either a measurement from a known location (i.e., a USGS benchmark) or calculation of location using a global positioning system (GPS). This known location becomes the permanent datum point from which all other spatial relationships are recorded.

The relationship of remains to each other and additional landmarks and evidence may be determined by the following means:

- Direct measurement from a known location

- A grid system (i.e., baseline measurements, which may or may not be situated along a cardinal direction and/or intersecting lines)

- Azimuth, distance, and inclination

- Vector, distance, and inclination, usually calculated from a known baseline (i.e., total station)

    —if baseline is north/south, it is an Azimuth, distance, and inclination

Buried remains are mapped by including depth information (i.e., the grid number and the depth below the grid strings, or azimuth, distance, and inclination from datum). Forensic archaeologists tailor their excavation towards finding contextual information about the grave. They may choose to pedestal the remains, remove the sidewall to search for shovel impressions, or enlist other archaeological techniques designed to record whatever information they deem important.

The first chore is to establish the datum point—a permanent point from which other measurements will be taken. From the datum, one can construct a baseline or set up a total station, transit, or other survey instruments. Once the parameters of the search area or grid are established, investigators begin to carefully remove all surface debris. If they find no remains on the surface, then they begin to excavate by removing the first centimeter of soil. Successive layers of soil are then removed, and all of the soil that is collected is screened through mesh to capture any small materials (especially teeth and small bones).

All remains are photographed in situ and mapped before they are moved. Once the lead investigator is satisfied that the scene has been completely and accurately documented, then the remains can be moved. The following rules apply to the movement of skeletal materials:

- Do not move the remains until every skeletal element has been photographed and its location measured and mapped.

- Move or lift the remains the shortest distance possible.

- Do not contaminate the scene while moving the biological remains.

- Place the skeleton or remains in a body bag, evidence bags, or a clean sheet, and secure them from prying eyes and undue handling.

- Place the hands of the decedent, if they are still articulated, in paper bags prior to moving the remains.

Processing Remains into Evidence

The general perception of forensic anthropologists is that they deal exclusively with skeletal remains. In fact, forensic anthropology laboratories receive, process, and analyze decomposed and burned remains, fragmented remains, and even various soft tissues (e.g., neck-organ blocks—see later chapters).

Initially the medical examiner or coroner should take custody of remains. For cases in which the forensic anthropologist has assisted in the recovery, arrangements may have been made to transport the remains directly to the forensic anthropology laboratory. Every time biological evidence moves from one place to another, both parties sign a receipt-of-evidence form. This receipt follows the biological evidence and becomes part of the permanent evidentiary record.

Bare Bones: A Survey of Forensic Anthropology

When remains arrive at the forensic anthropology laboratory, they are assigned a case number and entered into a **case logbook** that records all of the laboratory's cases. The logbook might record the medical examiner's case number for cross-referencing, the type of remains (e.g., skeletonized, decomposed, burned), whether the remains have been tentatively identified, and the place or agency of origin. The laboratory technician, manager, or director often uses a **processing log** that indicates what procedures are needed for that specific case. The processing log can be adjusted to the requirements of each case and records a timeline for each aspect of analysis and the general order of completion for each procedure. This allows the anthropologist to, years later, testify as to when each procedure was performed and by whom.

Each case also has an individual case file that will eventually include all forms and reports related to that case. The case file is a master record of each case, from receipt to release, and includes evidence receipts, shipping waybills, analysis forms, correspondence, photographs, radiographs, subpoenas and testimony transcripts, and the final forensic osteological report.

Maceration

If the remains are not completely skeletonized or are otherwise deemed to be biohazardous, then they are processed by a technique called maceration. Maceration is the process of removing any soft tissue from decomposing or mummified remains in order to yield a clean, dry skeleton for forensic anthropological analysis. Skeletal preparation and maceration is done in several ways, but the primary method involves immersing the skeleton into boiling or simmering water.

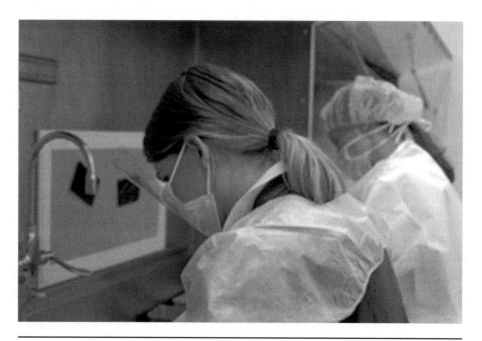

■**Figure 5:** *Katie Skorpinski and Laurel Freas [in background] macerating a case at the C.A. Pound Human Identification Laboratory*

The authors follow a traditional method by placing the remains in large stainless-steel pots. The pots are then filled with water and brought to boiling. Once the water is boiling, we reduce the temperature to slightly under 100 degrees Celsius. After several hours, we change the water, evaluate the bones, and repeat the process for up to two to three days, depending on the stage of decomposition. Our laboratory has had good success with this older method, which does not introduce any foreign substances to the bones, which might later affect chemical analyses (Walsh-Haney et al. 2008).

Once the soft tissue has softened, it can be removed from the underlying bone.

The soft tissue can, with minimal effort, be removed using nonmetallic tools such as plastic lab stirrers and spatulas or nylon brushes. The use of plastic and nylon tools prevents introduction of additional trauma (i.e., cut marks) to the bones that might complicate later analysis. In some cases the bones are

Bare Bones: A Survey of Forensic Anthropology

scrubbed with diluted (10:1) citrus-based degreaser and rinsed with clean water. Personnel are very careful to retain all possible soft tissue, which they place in evidence bags and freeze for storage. Once the skeletal remains are identified and/or returned to the medical examiner, the soft tissue is also returned. In some unidentified cases, tissues are retained for a period of years and then incinerated as biohazardous material.

Once maceration is complete, the remains are placed under a drying hood for three to seven days. The remains are now ready to be placed on the examination tables, in anatomical position, so that the examiner can take a thorough inventory. An inventory records exactly which bones and teeth staff received at the laboratory and helps insure that all remains are returned to the family if the remains are subsequently identified.

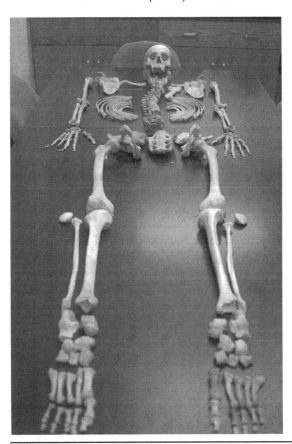

■FIGURE 6: *Remains in anatomical order on table in teaching lab*

Now that we have recovered the remains, entered them into evidence, and processed the remains until they are skeletonized, we can begin to discover the identity of the decedent and—if we are lucky—find evidence of how he or she died.

References

Dupras, T. L., J. J. Schultz, S. M. Wheeler, and L. J. Williams. 2006. *Forensic recovery of human remains*. Boca Raton: CRC Press.

Schultz, J. J. 2003. Detecting buried remains in Florida using ground-penetrating radar. PhD diss., Univ. of Florida.

—. 2008a. Sequential monitoring of burials containing small pig cadavers using ground-penetrating radar. *Journal of Forensic Sciences* 53 (2): 279–287.

—. 2008b. Using ground-penetrating radar to locate clandestine graves of homicide victims. *Homicide Studies* 11 (1): 15–29.

Walsh-Haney, H. A., L. E. Freas, and M. W. Warren, eds. 2008. The working forensic anthropology laboratory. In *The forensic anthropology laboratory*, 97–116. Boca Raton: CRC Press.

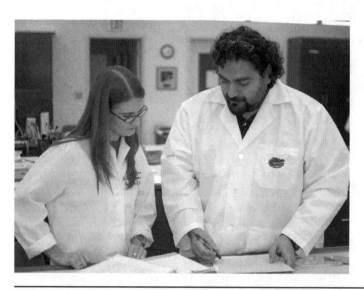

■FIGURE 7: *Katie Skorpinski and Carlos Zambrano review a peer-review checklist to make sure all policies and procedures have been followed during the the analysis of a case.*

Bare Bones: A Survey of Forensic Anthropology

Sample Test Questions

1. **Enclosed scenes include all the following EXCEPT:**

   a. Dumpsters

   b. Any scene in which the body is confined to a known area

   c. A clandestine grave

   d. Railroad cars

2. **List some of the goals of search and recovery.**

3. **Which method is an example of a noninvasive search method?**

   a. Grid searches

   b. Mapping the site

   c. Securing the scene

   d. Soil probing

4. **True or False:**

   Evidence for burial may be artifactual, environmental, geological, or a combination of these three categories.

5. **Soft tissue is removed from decomposing bodies by immersion in warm water, a process known as:**

   a. Defleshment

   b. Tissue deconstruction

   c. Dissection

   d. Maceration

Taphonomy

The term **taphonomy** comes from the Greek *taphos*, for burial, and *nomos*, for law, and literally translates into "the laws of burial" (Efremov 1940). The study of taphonomy began in an attempt to understand how fossils were formed. All vertebrates eventually pass from the biosphere to the lithosphere. Thus, the field of paleontology gave birth to taphonomy in order to study the processes that alter an organism after death, with particular emphasis on long-term changes (Nawrocki 1996).

Forensic taphonomy has developed as a subfield of forensic anthropology to analyze autolytic, decompositional, and environmental changes to the soft and hard tissues immediately after death. This newer approach has informed the paleontological study of fossil formation but has primarily been developed to answer several questions about the **postmortem interval (PMI)**, or time since death. The questions that are usually asked by the forensic anthropologist are as follows:

- How long has it been since the death of this individual?

- If the remains are scattered, how were they scattered?

- What lesions to the skeleton are the result of perimortem trauma, and what damage can be attributed to the postmortem period?

These questions and how the forensic anthropologist answers them will be discussed throughout the chapter.

TIME SINCE DEATH

Estimating the time since death of a decedent is a very important task and one of the first things asked by law-enforcement agencies. Most cases that go straight to the medical examiner's or coroner's office still have a significant amount of soft tissue. Thus, more traditional methods for determining time since death can be utilized. With fully fleshed bodies, there are three main ways to estimate the amount of time that has elapsed since an individual's death: rigor mortis, livor mortis, and algor mortis.

Rigor mortis is the amount of muscle stiffness in a body after the time of death. The settling of blood in the body is referred to as **livor mortis.** This is helpful in estimating not only the time since death but also the position of the body at the time of death, since the blood tends to pool within the lowest parts of the body. **Algor mortis** is the cooling of the body temperature after the time of death. Each of these indicators is useful for up to twenty-four hours after death. However, after this time period, these indicators begin to lose accuracy, and the medical examiner is forced to turn to other methods for estimating time since death. Thus, the medical examiner turns to the forensic anthropologist or for an analysis of taphonomic indicators. It should also be noted that forensic entomologists—experts in necrophagus insects—can usually offer a better estimate of the postmortem interval than can anthropologists. Extensive research into the life cycles of insects known to colonize dead bodies enables the entomologist to collect and identify the various species inhabiting a body and establish the age of successive larval and adult stages of each. Insects, along with autolysis of tissues by normal and invasive florae of bacteria, play perhaps the most important role in the early decomposition of living organisms.

There has been much research on time since death and decomposition rates of the human body in many areas of the country. However, it is important to note that each study is specific to the geographical area where the study was conducted. The most notable studies include three geographical areas and different environments, each of which gives a detailed description of the rate of decomposition and skeletonization. Bass (1997) describes decomposition rates in warm, moist climates; Galloway and colleagues (1989) evaluate the changes in the body after death in warm, dry climates; and lastly, Komar (1998) illustrates decomposition rates in a cold environment.

The research provided by colleagues in these general geographic zones has been very informative for practitioners in each region. However, much of what has been learned is applied in terms of how, through the experience of anthropologists, remains found within their areas of practice vary from the abovementioned centers of taphonomic research. No two environments are exactly the same. Many variables may affect decomposition and scavenging rates. For example, in subtropical Florida, microenvironments produce dramatically different decomposition rates depending on direct exposure to sunlight, humidity, local populations of necrophagus insects, and other factors.

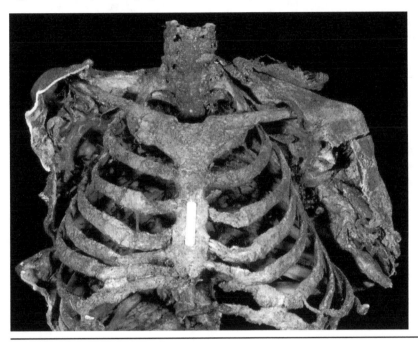

■FIGURE 1: *"Wet" mummification*

SCATTER OF REMAINS

The majority of the skeletal remains recovered by a forensic anthropologist are surface scatters, not burials. The extent of the scatter of remains is variable; however, often skeletal elements are scattered to a much wider extent than one might think. Skeletal remains can be scattered either intentionally or unintentionally through a variety of ways, including natural scatter, animal predation, human scatter, or a combination of more than one agent.

Natural Scatter

As the soft tissue of the body begins to undergo the process of decomposition, the joints will deteriorate in a systematic manner. The weakest joints (**synovial joints**) are the first to **disarticulate**, or come apart. These are also the most common joints found in the human body. For example, the elbow joint formed by the humerus, ulna, and radius is often one of the first to disarticulate. The next joints to decompose are the joints of intermediate strength (**cartilaginous joints**). The articulations of the vertebrae are an example of this type of joint. The last joints to disarticulate are the strongest ones (**fibrous joints**). The union of the two sides of the pubic symphysis is one of the strongest joints in the human body and usually one of the last ones to decompose. Additionally, sutures between cranial bones typically hold fast due to their interdigitation. Once the body is fully skeletonized, it will begin to disperse due to natural processes such as animal predation and weather scatter.

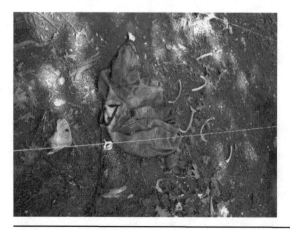

■FIGURE 2: *Surface deposition.*

Bare Bones: A Survey of Forensic Anthropology

Animal Predation

The two most common types of animal predation seen in forensic anthropology are carnivore modification and rodent modification. In addition to scattering remains, animals can also modify the remains by permanently removing them from the scene through ingestion of parts of entire bony elements (Haglund et al. 1988).

Carnivore Modification

Carnivores typically begin to modify the body soon after the time of death. A dead human body may be viewed as a nutritional boon to large predators. Many carnivores will feed on carrion. Dogs and coyotes are the most frequent carnivorous predators among human remains and can leave a variety of marks on the bone (Haglund et al. 1988). Puncture marks appear as round or oval-shaped indentions in the bone caused by the canines of the carnivore (Figure 3: Carnivore marks). Carnivores can also leave crushing fractures due to chewing on the bone with the cheek teeth. Lastly, gnaw marks in the appearance of grooves or scratches can also be seen on the bone. Puncture marks and crushing fractures are more common on the ends of long bones, while gnaw marks are found primarily on the shaft of the bone. In cross section, carnivore scoring marks tend to be more v-shaped in comparison to rodent gnaw marks. Movement of bones, especially of the skull, may represent play behavior in pets and feral dogs. It is not uncommon for remains to be located after the family pet brings a human bone home. Coyotes may remove bones and other tissues and take them back to their dens. In areas where coyotes are present, a thorough search is required to find the dens and recover the remains.

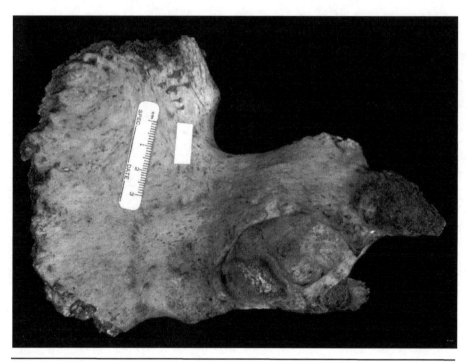

■FIGURE 3: *Carnivore marks.*

RODENT MODIFICATION

Unlike carnivores, rodents do not begin to modify the body until the majority of the soft tissue has decomposed or been removed. Rodents have a specialized chewing apparatus with two incisors that never stop growing. As such, it is necessary for them to continuously file down their teeth to ensure that they do not grow too long. The process of rodent modification of bone is similar to what many individuals may have seen in the processing of wood by rodents. Like wood, bone also makes the perfect substance to file down the teeth. Bones also provide essential nutrients to the rodent's diet. Squirrels have been known to carry smaller bones up trees so that they can safely gnaw on them. In wooded areas, it is best to search near the bases of trees for smaller bony elements.

Rodent activity on bone can be differentiated from carnivore activity by the characteristic parallel **striae** left on the bone from the rodents' unique incisors. Rodents primarily will target the epiphyses of the long bones, followed by the adjacent shafts (Haglund 1992). Rodent modification can be differentiated from carnivore scoring marks, as rodent gnaw marks are more u-shaped in cross section.

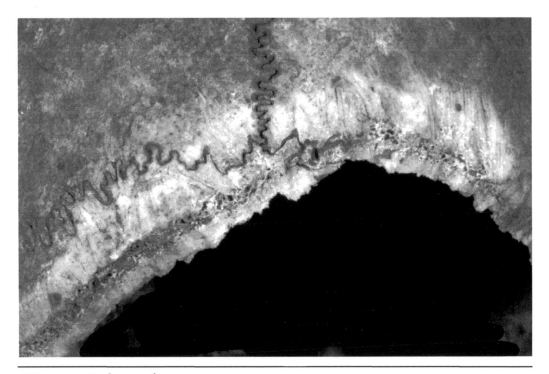

■FIGURE 4: *Rodent marks*

Human Scatter

In addition to carnivore and rodent scatter of skeletal remains, humans themselves may also play a part in the dispersal of remains, either unintentionally or intentionally.

Unintentional

Unintentional forms of scattering remains can occur in situations where a body has been deposited in a particular location, unbeknownst to the person scattering the remains. A frequent case of unknown scattering occurs when a body has been deposited in a farming plot and remains undiscovered during farming operations. In this type of a case, damage to the bone from the farming equipment is evident and thus attributed to unintentional postmortem damage.

INTENTIONAL

Dismemberment is the act of cutting up, or dissecting, the body in order to facilitate transport of the remains, hide the remains, or make identification of the remains more difficult for medicolegal investigators. Additionally, determination of cause of death is also obscured, since the traumatic marks made from the cutting instrument may obliterate other signs of trauma that occurred around the time of death (Di Nunno 2006). Almost all dismemberment cases exhibit saw marks. In many cases, there is evidence that a knife was used to cut through the soft tissues, and then a saw was used to cut through the bones. Saw marks due to dismemberment are discussed in more detail in a following chapter on trauma analysis.

Natural and animal disarticulation can be differentiated from human dismemberment by careful examination of the areas near the joint spaces. Natural disarticulation occurs around the joint surfaces with no evidence of saw marks, whereas the majority of human disarticulations occur in the proximal and distal areas of the bones, leaving clear evidence of the method of dismemberment.

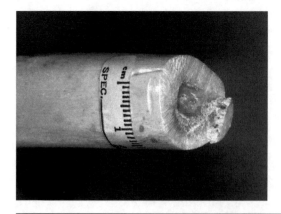

■FIGURE 5: *Note the striae capturing the individual cuts of the saw's teeth.*

Bare Bones: A Survey of Forensic Anthropology

Another type of intentional damage to human remains occurs when the perpetrator attempts to burn a body to get rid of evidence. In such a case, the damage from the fire will dramatically alter the bone, making it more difficult—but not impossible—to identify the individual. The fire damage may also obscure evidence of perimortem trauma. In some instances, after a body has been burned, the perpetrator may then decide to scatter the remains over a large area to further obstruct law enforcement.

Environmental Processes on Bone

Almost any environmental process can alter the appearance of bone. Among the most active taphonomic agents cited are weathering, water, heat, soils, and plant life.

Weathering

The appearance of bone can change dramatically with exposure to natural weathering processes such as rain, heat, humidity, cold, etc. Once the body is devoid of soft tissue, the skeleton can be affected by any of these environmental processes. **Delamination** of bone occurs when the outer layer of the cortical bone begins to peel away from the underlying layers. The cracking of the cortical bone follows the internal bone-fiber structure. This process is typical in bone that has been exposed to a variety of environmental factors (Behrensmeyer 1978). The sun also has the ability to change the appearance of bone. When the skeleton has been exposed to the sun for a long period of time, it tends to take on a white appearance, known as **sun bleaching**, shown in figure 6 (Ubelaker 1999). This white color can be differentiated from the white of calcined bone, as the bone still has much of its organic component to it. In addition to mere color change, sun bleaching will also cause the outer cortex of the bone to become flaky, with small cracks along the surface.

There are several other factors that may affect the appearance of bone long after the time of death of an individual. **Soil staining** on the skeleton occurs when the bone has been in direct contact with soil. In such cases, the color of the soil affects the color of the bone. For example, in clay-rich areas, bone may take on a reddish appearance due to close contact with the clay. In many instances plant vegetation may grow in, around, or on top of skeletal remains. Acid secreted by roots will literally etch into the surface of the bone, leaving characteristic **root-etching marks** on the bone (Gill-King 1997).

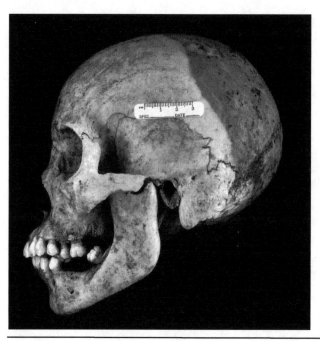

■FIGURE 6: *Note the contrasting color pattern on the skull. The dark staining is a result of contact with the soil, which indicates the position in which it sat after decomposition of the body.*

Water Transport

The movement of human remains through water is known as **fluvial transport.** Remains that have been deposited in water, such as a river or creek, display indicators typical of this aqueous habitat. One characteristic is perforation of bones caused by exposure to small rocks or pebbles in the water, leading to damage of the thin bones of the face. Additionally, these rocks may create abrasion or pitting on the bones where the surface of cortical bone is damaged, giving it a "pitted" appearance much like the surface of the moon. Lastly, microenvironments with shade and moisture may stimulate growth of algae or fungi on skeletal elements, giving the bone a greenish hue (Nawrocki et al. 1997).

Heat and Fire Damage

Damage to the skeleton from fire warrants special consideration and is addressed by the authors in a later chapter. In brief, bone that is taphonomically altered by intense heat or fire has a very distinct appearance and is

Bare Bones: A Survey of Forensic Anthropology

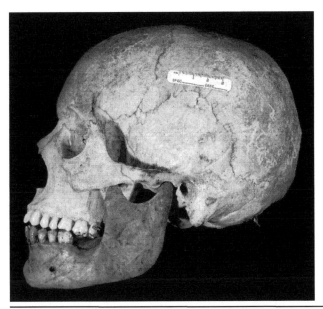

■FIGURE 7: *Note the different colors of the cranium and mandible. The mandible can become disarticulated after decomposition, and be exposed to slightly different taphonomic processes.*

unlikely to be confused with any other type of damage. Fire damage to bone will not only change the color of the bone, but it also may change its shape. The color change of burned bone progresses systematically as more heat is applied and time elapses. The typical color changes from least burned to most burned is as follows: yellow ➤ light brown ➤ black ➤ blue-gray ➤ white. Charred bone refers to bone with a black coloration. Once the organic component of bone has been removed and only the inorganic component remains, the bone is often referred to as **calcined**, which refers to the stage of whitish-gray coloration.

In addition to color changes, bone that has been subjected to intense heat will sometimes undergo nonreversible shape changes. In particular, bone is known to sometimes shrink up to 20 percent of its original size. This process is sometimes referred to as bone shrinkage. Another modification to shape is **warping**, where the bone shape changes from its original architecture. Both shrinkage and warping are important factors to consider, as measurements of bones exposed to intense heat will not accurately reflect the true dimensions of the bone. As such, determination of many aspects of the biological profile, such as sex, ancestry, and stature, will be affected if using calculations taken from burned bone (Correia 1997a, 1997b). The effect of shrinkage is magnified during the fetal, neonatal, and infancy periods, since similar degrees of shrinkage create a relatively larger error (Huxley and Kósa 1999).

Perimortem versus Postmortem Trauma

The anthropologist must be able to differentiate perimortem trauma that occurred at the time of death and postmortem damage that occurred long after the time of death. Trauma is identified in two ways: (1) the appearance and characteristics of the fractured margins of bone, and (2) the likelihood that a given fracture was produced by a mechanism seen in clinical medicine.

Perimortem trauma that occurs on green, or fresh bone, has a very different appearance than postmortem damage on dry bone. When bone is broken at or around the time of death, the margins of the fracture are smooth. Also, the fractures occur as transverse, oblique, spiral, or comminuted types at areas of the bone that are vulnerable to outside forces. Fractures must make sense from a clinical perspective. Do living people break their bones in this way? If not, then the anthropologist must consider that the break in the bone occurred after the remains were skeletonized and therefore subject to different forces. Blunt-force traumatic injury is discussed in detail in a following chapter on trauma.

Postmortem breakage leaves the edges of the bone jagged and irregular. The orientation of fractures on green bone will spiral along the shaft of the bone, whereas postmortem fractures are oriented transversely or perpendicular to the shaft of the long bone. Lastly, the color of the fracture can also be a good indicator of a fracture that occurred at or after the time of death. If the color of the exposed bone along a line of fracture is different than that of surrounding bone, it is most likely due to postmortem damage.

Bare Bones: A Survey of Forensic Anthropology

References

Bass, W. M. 1997. Outdoor decomposition rates in Tennessee. In Haglund and Sorg 1997.

Behrensmeyer, A. K. 1978. Taphonomic and ecologic information from bone weathering. *Paleobiology* 4:150–162.

Correia, P. M. M. 1997a. Fire modification of bone: A review of the literature. In Haglund and Sorg 1997.

Correia, P. M. M., and O. Beattie. 1997b. A critical look at methods for recovering, evaluating, and interpreting cremated human remains. In *Advances in forensic taphonomy: Method, theory, and archaeological perspectives*, ed. W. D. Haglund and M. H. Sorg, 435–450. Boca Raton: CRC Press.

Di Nunno, N., F. Constantinides, M. Vacca, and C. Di Nunno. 2006.

Dismemberment: A review of the literature and description of 3 cases. *American Journal of Forensic Medicine and Pathology* 27:307–312.

Efremov, I. A. 1940. Taphonomy, a new branch of paleontology. *Pan-American Geologist* 74:81–93.

Galloway, A., W. H. Birkby, A. M. Jones, T. E. Henry, and B. O.

Parks. 1989. Decay rates of human remains in an arid environment. *Journal of Forensic Sciences* 34:607–616.

Gill-King, H. 1997. Chemical and ultrastructural aspects of decomposition. In Haglund and Sorg 1997.

Haglund, W.D. 1992. Contribution of rodents to postmortem artifacts of bone and soft tissue. *Journal of Forensic Sciences* 37:1459–1465.

Haglund, W. D., D. T. Reay, and D. R. Swindler. 1988. Tooth mark artifacts and survival of bones in animal scavenged human skeletons. *Journal of Forensic Sciences* 33:985–997.

Haglund, W.D., and M. H. Sorg, eds. 1997. *Forensic taphonomy: The postmortem fate of human remains*. Boca Raton: CRC Press.

Huxley, A. K., and F. Kósa. 1999. Calculation of percent shrinkage in human fetal diaphyseal lengths from fresh bone to carbonized and calcined bone using Petersohn and Köhler's data. Journal of Forensic Sciences 44 (3): 577–583.

Komar, D. A. 1998. Decay rates in a cold climate region: A review of cases involving advanced decomposition from the medical examiner's office in Edmonton, Alberta. *Journal of Forensic Sciences* 43:57–61.

Nawrocki, S. 1996. An outline of forensic taphonomy. University of Indianapolis Archeology and Forensics Laboratory. http://archlab.uindy.edu/documents/ForensicAnthro.pdf.

Nawrocki, S., J. Pless, D. Hawley, and S. Wagnes. 1997. Fluvial transport of human crania. In Haglund and Sorg 1997.

Ubelaker, D. H. 1999. *Human skeletal remains: Excavation, analysis, interpretation.* Washington, D.C.: Taraxacum Press.

Sample Test Questions

1. **Rigor mortis is best defined as:**

    a. The period of time since death

    b. The amount of muscle stiffness in a body after the time of death

    c. The cooling of the body temperature after the time of death

    d. Settling of blood in the body

2. **Which joints are the first to disarticulate?**

    a. Fibrous joints

    b. Cartilaginous joints

    c. Epiphyses

    d. Synovial joints

3. **All of the following are examples of different types of weathering EXCEPT:**

    a. Warping

    b. Delaminating

    c. Sun bleaching

    d. Root-etching marks

4. **Which of the following color changes represents the least amount of fire damage?**

    a. White

    b. Black

    c. Blue-grey

    d. A and B only

    e. All of the above

5. **Which of the following color changes represents the most amount of fire damage?**

   a. White

   b. Black

   c. Blue-grey

   d. A and B only

   e. None of the above

Human Osteology and Osteometry

Osteology is the study of bones. Why study bones? Well, they last a long time, thereby providing biological evidence for early humans and proto-humans. Bones enable us to better understand ancient populations, their patterns of growth and development, anatomical adaptations and overall health. And, as will be seen in this book, evidence gleaned from the human skeleton can help us solve certain types of forensic cases.

The skeleton is a complex system. The proper study of osteology requires intense study over a period of years. However, it is necessary for the reader of this volume to have at least a basic appreciation for the skeleton to understand the concepts, methods and techniques used by forensic anthropologists in their jobs. Before we begin our tour of the human skeleton, we must first orient ourselves using some basic anatomical and medical terminology.

ANATOMICAL PLANES, POSITIONS AND DIRECTIONS

Three primary planes are used in dividing the skeleton. The **sagittal plane** passed through the midline of the body, dividing it into equal left and right halves. Any planar slice in this direction that is not midline is called para-sagittal. The **coronal plane** divides the skeleton into equal anterior and posterior sections. The **transverse plane** cuts the body into upper and lower sections. This has become a very important plane in clinical medicine because of diagnostic techniques such as computed axial tomography scanning and magnetic resonance imaging, which presents the clinician with "slices" of transverse anatomy.

Since the head is mobile, scientists need a way to standardize head position so that measurements are comparable. The answer was found during an early anatomical conference held in Frankfort, Germany. After considering the posture of a German soldier at attention and discussing the merits of several suggestions, the anatomists arrived at the Frankfort Horizontal Plane – a position that aligns the lower border of the left eye orbit (the osteometric point *orbitale*) with the superior aspect of the external ear openings (the osteometric point *porion*). This position is best demonstrated by the use of a Case-Western head spanner and craniophor (Figure 1).

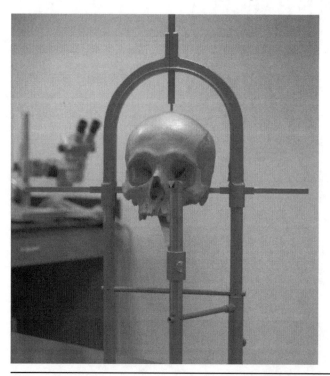

■**FIGURE 1:** *This cranium is held by a head spanner, or craniophor, allowing for measurement of vertex and mastoid length.*

Bare Bones: A Survey of Forensic Anthropology

Anatomical position in humans is unique due to our bipedal form of locomotion. Structures that would be described as superior and inferior in quadrupeds become anterior and posterior in humans. Correct anatomical position for humans is standing, with feet pointing forward, arms by the sides, and with the palms of the hands facing forward and the thumbs pointing away from the body. This seems like an unnatural position for some people, particularly robust males who normally carry their hands rotated inwardly. This relaxed posture is achieved by pronating the forearm, resulting in the radius crossing the ulna. Anatomical position keeps the radius lateral to the ulna and in its proper position as the forelimb counterpart to the fibula.

Osteologists must describe landmarks on bones as they relate to other structures. For example, the greater tubercle of the humerus is *lateral* to the bicipital groove, or, the gunshot wound is *anterior* to the coronal suture. Table 1 below lists some anatomical directions commonly used by anthropologists.

TABLE 1: *Anatomical directions.*

superior/cranial	toward the top
inferior/caudal	toward the bottom
medial	toward the midline
lateral	away from the midline
anterior/ventral	toward the front
posterior/dorsal	toward the back
proximal	nearest the trunk
distal	farthest from the trunk

The Skeleton

The bones of the skeleton can be categorized in a number of ways. Bones may be classified by the way in which they develop (either *intramembranous* or *endochondral*), their shape (*e.g., long bones, flat bones*, or *irregular bones*), or whether they comprise part of the skull or body (*cranial* or *postcranial*). In this chapter, we will divide the skeleton into those elements that lie along the axis of the body (*axial*) or constitute part of the limbs (*appendicular*).

The Axial Skeleton

The axial skeleton includes all of the bones of the skull and the vertebral column. The skull consists of the cranium, mandible and ear ossicles. For our purposes, we will also include the hyoid bone, which is situated inferior to the mandible and suspended from the styloid process of the temporal bone. The vertebrae normally consist of 33 bones, divided into the cervical, thoracic, lumbar, sacral and coccygeal sections.

The skull consists of 29 bones. Six bones are unpaired – they lie along the midline. Eleven are paired into left and right elements (**Table 2**). The final bone is the aforementioned hyoid. **Figures 2 through 4** show three views of the skull with the major bones labeled. (Note: All of the following photographs in Chapter 5 are courtesy of Merissa Olmer)

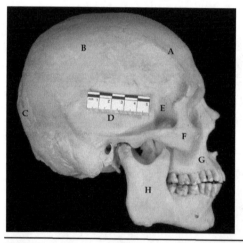

■Figure 2.

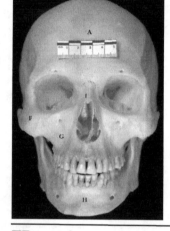

■Figure 3.

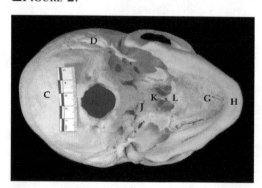

■Figure 4.

Bare Bones: A Survey of Forensic Anthropology

TABLE 2: *The bones of the skull.*

BONE	PAIRED OR UNPAIRED
Frontal (A)	unpaired
Parietal (B)	paired
Occipital (C)	unpaired
Temporal (D)	paired
Sphenoid (E, J)	unpaired
Zygomatic (F)	paired
Maxilla (G)	paired
Mandible (H)	unpaired
Nasal (I)	paired
Vomer (K)	unpaired
Palatine (L)	paired
Ethmoid (not shown)	unpaired
Lacrimal (not shown)	paired
Stapes (not shown)	paired
Incus (not shown)	paired
Malleus (not shown)	paired
Hyoid (not shown)	unpaired
Inferior Nasal Concha (not shown)	paired

Anatomists divide the vertebral column in segments. The vertebrae from each segment have distinguishing characteristics related to their function and location along the vertebral column. Additionally, some vertebra are unique among those within their segment, so osteologists are able to identify an isolated vertebra with a specific number within the segment (*e.g.*, a *12th* thoracic vertebra).

The cervical vertebrae are in the neck. This segment shows the least variation among and within mammalian species, almost *always* having 7 vertebra. Giraffes, humans and mice all have 7 cervical vertebra. Among the cervical vertebra, numbers 1 and 2 differ from all other vertebrae in the column. The first cervical vertebra is called the *atlas* – named after the Greek god of heavy burdens, Atlas, who holds the world upon his shoulders. Like Atlas, the first cervical vertebra supports our head. The atlas is the only vertebra without a body, or centrum. Instead, it has an anterior arch that articulates with the odontoid process of the second cervical vertebra (**Figure 5**). The second cervical vertebra is called the *axis*. This vertebra has an extension on the superior aspect of its centrum called the odontoid process, or dens. This odontoid process passes behind the anterior arch of the atlas, which rotates around the axis of the dens (**Figures 6 and 7**).

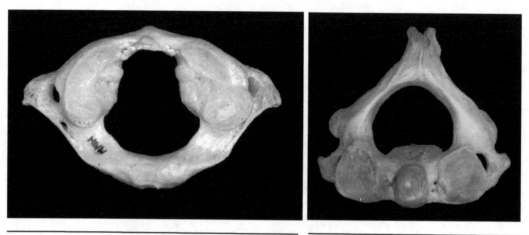

■FIGURE 5.   ■FIGURE 6.

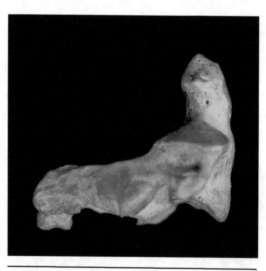

■FIGURE 7.

Bare Bones: A Survey of Forensic Anthropology

The remaining cervical vertebra share common features. The single characteristic shared by all cervical vertebrae is the transverse foramen, through which passes the cervical artery (**Figure 8**).

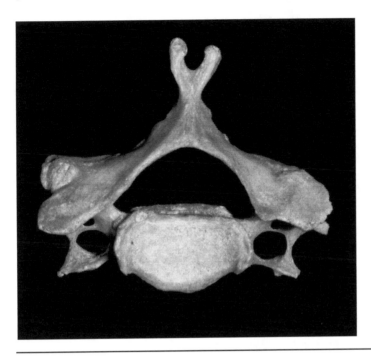

■FIGURE 8.

The twelve thoracic vertebrae lie within the thorax, or chest. The ribs attach to the centrum of each vertebra and, through cartilaginous tissue, to the sternum at their distal ends. The thorax acts as a mechanical bellows, expanding and contracting to move air in and out of the lungs. So, a common feature shared among all thoracic vertebrae is a costal facet – the articulation for the rib. The first, tenth, eleventh, and twelfth vertebral centra each have a single, or whole costal facet. The remaining thoracic vertebrae have demi-facets, since the ribs in this portion of the chest articulate between vertebral centra. Therefore, each of these vertebra share one-half of a rib articulation with their superior and inferior neighbors (**Figure 9**).

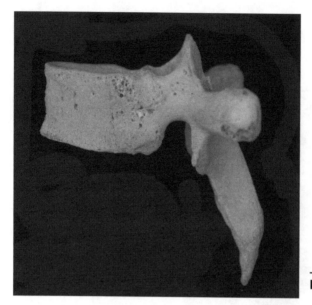

The five lumbar vertebra are the largest. They lack a transverse foramina of the cervical vertebrae and the costal facets of the thoracic vertebrae. Their superior and inferior articular facets are concave and convex, respectively, allowing for rotation of the column (**Figure 10**).

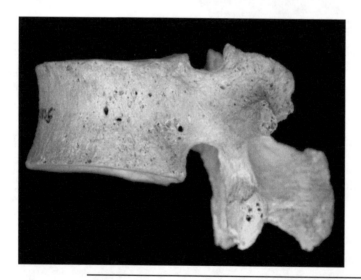

■FIGURE 10.

Bare Bones: A Survey of Forensic Anthropology

The sacrum consists of five vertebral bones that fuse to form one element. The sacrum makes up part of the pelvic girdle. It articulates with the os coxa, or hip bones, at the sacroiliac joint (**Figure 11**). The coccyx consists of 4 vertebral bones that represent remnant caudal vertebrae . . . our tails!

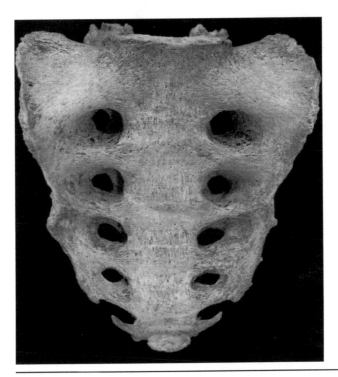

■FIGURE 11.

The ribs attach to the vertebral centra on the lateral aspect anterior to the transverse processes. The upper ten ribs also articulate with the transverse processes at the costal foveae. The ribs protect the vital structures of the thorax, and as mentioned above, operate as a bellows to move a tidal volume of air in and out of the lungs. Each rib has a head, neck, and shaft (**Figure 12**).

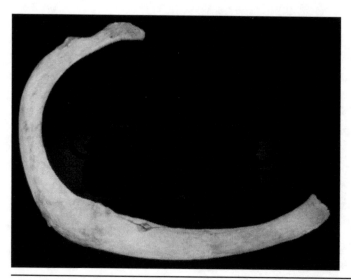

■FIGURE 12.

The 1st, 2nd, 10th, 11th and 12th ribs are identifiable based on unique morphology.

The ribs articulate with the sternum anteriorly, via the costal cartilage. The top 6 or 7 ribs each have their own costal cartilage and are called **true ribs**. The remaining ribs share costal cartilage and are called **false ribs**. The bottom two sets of ribs are known as the **floating ribs**, since they do not articulate with the transverse process of the thoracic vertebra and do not extend to the sternum.

The sternum, or breastbone, consists of three elements: the manubrium, the body, and the xiphoid process. The sternum consists of fused sternebrae in the adult. The 1st sternabra is the manubrium, and the 2nd through 5th sternabrae fuse in most adults to make up the body. The manubrium is the most superior of the elements. It can be palpated at the "U-shaped" jugular notch between the clavicular notches along its superior border. The body, or corpus sterni, is the blade-like main portion of the sternum (**Figure 13**). The body has notches along its lateral border to accept the articulations of the costal cartilage of the upper ribs. The xiphoid process lies at the most inferior aspect of the body of the sternum. It is cartilaginous, but often ossifies in older individuals.

Bare Bones: A Survey of Forensic Anthropology

■Figure 13.

The Appendicular Skeleton

The appendicular skeleton consists of the bones that contribute to the extremities, and so include the bones of the shoulder girdle, pelvic girdle, arms, hands, legs and feet.

The shoulder girdle

The shoulder girdle is comprised of the clavicle and the scapula. The clavicle, or collarbone, is located on the superior aspect of the thorax. It articulates with the manubrium of the sternum medially, and the acromion process of the scapula laterally. The clavicle is the first skeletal element to ossify during development, and its epiphysis is the last to fuse to the primary center. It is also the only bone of the postcranium that develops within membranes (membranous or dermal bone). The remainder of the postcranial skeleton forms within a cartilaginous matrix (endochondral bone). The landmarks of the clavicle are shown in **Figure 14** below.

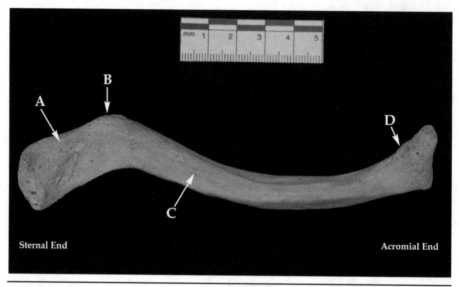

Sternal End

Acromial End

■**FIGURE 14:** A, acromial end; B, conoid tubercle; C, subclavian sulcus; D, costal impression.

The scapula, or shoulder blade, is a large, flat, triangular bone lying on the posterior aspect of the thorax. The scapula articulates with the humerus at the glenoid fossa and with the clavicle at the acromion process. The scapula is free to move about to allow range of motion of the upper extremity. It is encased within the thick muscles of the rotator cuff. The triangular shape has three borders: the superior border, the axillary (or lateral) border, and the vertebral (or medial) border. The scapula is shown below in **Figure 15**.

Bare Bones: A Survey of Forensic Anthropology

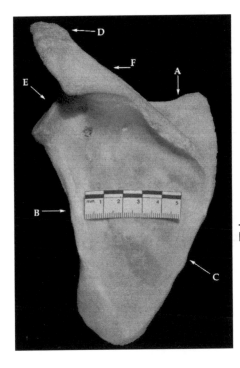

■FIGURE 15: *A, superior border;*
B, axillary border;
C, vertebral border;
D, acromian process;
E, glenoid fossa;
F, spine.

Long bones

All long bones share a similar architecture. The shaft of the bone is called the **diaphysis**. The long bones are endochondral bones, that is, they ossify within a cartilaginous matrix. The diaphysis ossifies as a primary center, beginning with a "bone collar" that becomes invaginated by a nutrient blood vessel. Once the diaphysis has ossified, it begins to elongate at either end. The end of the diaphysis is called the **metaphysis**. Bone growth takes place along a growth plate called a **physis**, which is situated at the ends of the metaphyses. All growth takes place on the metaphyseal side of the growth plate. A secondary ossification center, the epiphysis, later begins to appear. When growth ceases, this epiphysis fuses to the metaphysis and the bone has reached its adult form (**Figure 16**).

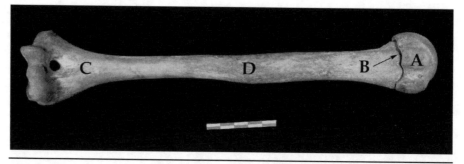

FIGURE 16: *A, epiphysis; B, physis, or growth plate; C, metaphysis; D, diaphysis.*

The arm

The humerus is the bone of the upper arm and the largest bone in the upper limbs. The humerus articulates with the glenoid fossa of the scapula on the proximal end, and the radius and ulna at the distal end. The proximal joint is a ball and socket, so the humerus has a ball, or head, at its proximal end. The distal end has articular surfaces to accommodate the two bones of the forearm. The humerus is shown in **Figure 17**.

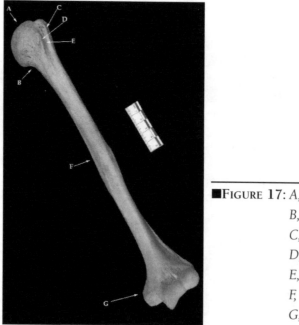

FIGURE 17: *A, head;*

B, neck;

C, lesser tubercle;

D, greater tubercle;

E, intertubercular groove;

F, shaft;

G, medial epicondyle.

Bare Bones: A Survey of Forensic Anthropology

The forearm

The forearm bones are the radius and ulna. The ulna provides for a stable elbow joint, permitting flexion and extension at the elbow joint but limiting lateral movement. The radius enables one to pronate or supinate the hand and wrist, and provides almost all of the articulation between the forearm and wrist bones. Photographs of the radius and ulna are shown below in **Figures 18a and 18b.**

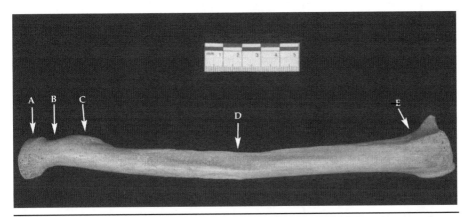

■FIGURE 18A: *A, head; B, neck; C, radial tuberosity; D, shaft; E, ulnar notch.*

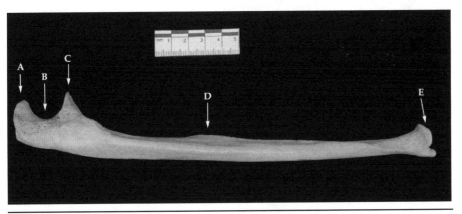

■FIGURE 18B: *A, oleacranon process; B, trochlear, or semilunar notch; C, coronoid process; D, shaft; E, radial articulation.*

The wrist and hand

The wrist and hand has 27 bones, not including a variable number of sesamoid bones. The eight bones of the wrist are called **carpals**. Each has a distinctive shape and can easily be identified by an experienced osteologist. Many osteology instructors honor a long tradition by having their students learn to identify each bone by touch alone – placing a single carpal into a black bag and having the student feel the size and shape of the bone. The carpals have complex articulations, so it is often difficult for students to remember the names of the bones by examining their articular surfaces. Many students overcome this problem by trying to identify the shape of each carpal with an object or animal. The capitate, for instance, has been described as having the appearance of the Star Wars character, Darth Vader, when held a particular way.

Each hand has five digital rays. The five bones of the palm are called **metacarpals**. The individual metacarpals are best identified by examining their bases. Each finger has three bones called **phalanges**, with the exception of the thumb, which has been reduced through evolution to allow it to oppose the other fingers and permit brachiation (**Figure 19**).

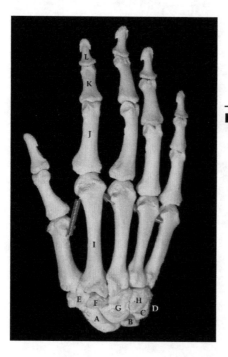

FIGURE 19: *A, scaphoid;*

B, lunate;

C, triquetral;

D, pisiform (not visible);

E, trapezium;

F, trapezoid;

G, capitate;

H, hamate;

I, metacarpal;

J, proximal phalange;

K, intermediate phalange;

L, distal phalange.

Bare Bones: A Survey of Forensic Anthropology

The leg

The femur, or thighbone, is the largest bone in the skeleton and articulates with the os coxae to form the hip joint. This joint is a ball and socket joint similar to the humerus, thus has a head and neck on its proximal end. The distal end articulates with the bones of the lower leg, the tibia and fibula, at the knee joint. As will be seen in later chapters, the femur has great utility as a diagnostic tool for forensic anthropologists. Since it contributes the most among the long bones to an individual's height, it the best bone to measure for estimating the living stature of the decedent. The head of the femur is also sexually dimorphic, being significantly larger in diameter among males than females, and so has utility in determining the biological sex of the individual. Finally, the femur shape varies among widely dispersed populations and can be used, in some instances, as an indicator of ancestry. The femur and its landmarks are shown below in **Figure 20**.

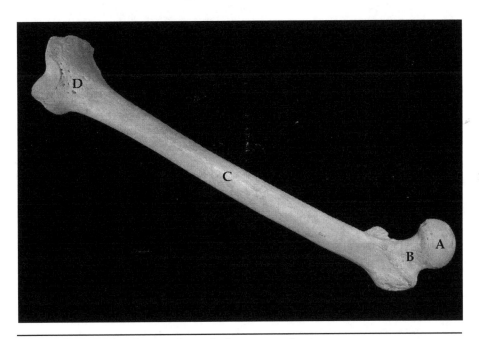

■**FIGURE 20**: *A, head; B, neck; C, shaft; D, metaphysis.*

The lower leg

The bones of the lower leg are the tibia and fibula. The proximal end of the tibia articulates with the condyles of the femur at the knee. All of an individual's weight is borne upon the tibial plateau, the flattened articular surface of the proximal tibia. The patella, or kneecap, is situated anterior to the knee joint. This is the largest of the sesamoid bones, forming within the quadriceps and patellar tendons of the knee extensor muscles.

The fibula articulates below and at the lateral aspect of the tibial plateau and bears no direct weight. The distal end of the fibula extends beyond the distal end of the tibia, and with the tibia, produces a mortise and tenon joint that provides lateral support for the ankle joint. The various landmarks for the tibia and fibula are illustrated below in **Figures 21**.

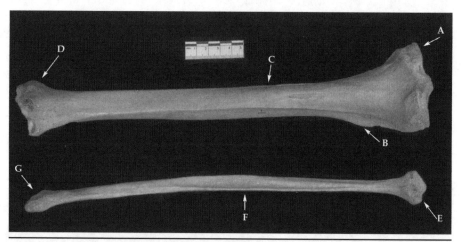

■**Figure 21**: *Posterior view of tibia. A, plateau; B, oblique view of tibial tuberosity; C, shaft and interosseous crest; D, articulation for fibula on the tibial pilon; E, fibular head; F, shaft; G, articulation with tibia.*

The ankle and foot

There are 26 bones in each ankle and foot. The seven bones of the foot are called the **tarsals**. Each is irregularly shaped and, like the carpals in the wrist, shares multiple articular surfaces with neighboring bones. Each foot has five **metatarsals**, representing the five digital rays. The toes each have a proximal, intermediate, and distal phalange, with the exception of the large toe, or **hallux**, which only has a proximal and distal phalange. The metatarsals and phalanges of the foot are different in cross-section from the metacarpals and

Bare Bones: A Survey of Forensic Anthropology

phalanges of the hand, in that they are more oval. The oval cross-sectional shape is better able to withstand the forces of bipedal locomotion than the "D-shaped" bones of the hand. The ankle and foot are shown in **Figure 22**.

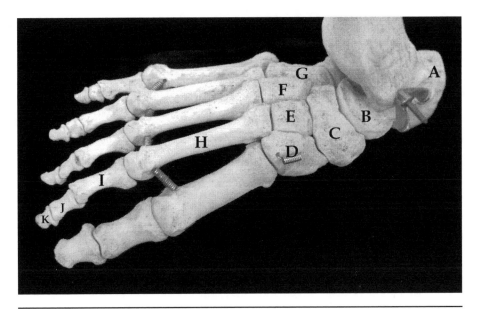

■FIGURE 22: *A, calcaneus; B, talus; C, navicular; D, 1st cuneiform; E, 2nd cuneiform; F, 3rd cuneiform; G, cuboid; H, metatarsal; I, proximal phalange; J, intermediate phalange; K, distal phalange.*

OSTEOMETRY

Osteometry is the study and measurement of the human skeleton. The measurements are used to compare and contrast the morphology and size of different populations, calculate stature, determine sex and ancestry, and in general, quantify whatever aspect of the skeleton that is being investigated. Anthropologists measure the skeleton according to *convention*, that is, scientists have mutually agreed upon standard points and techniques so that the data are comparable between researchers. Studies utilizing osteometric techniques provide detailed descriptions of the techniques and osteometric points used so that other researchers are clear about how the data were collected. Many of the osteometric points used today were defined over one hundred years ago (**Figure 23**).

■**FIGURE 23**: *Anthropologist Carlos Zambrano measuring a vertebra for metric analysis.*

Osteometric points may be anatomically-defined or instrumentally-determined. For example the osteometric point **bregma** is defined as "the point where the sagittal and coronal sutures meet" (Moore-Jansen *et al*., 1994). The osteometric point **euryon**, however, can be determined only by finding that point on the parietal bones that represent the widest point of the cranial vault.

Anthropologists primarily use two types of measuring calipers. Spreading calipers have outwardly arcing arms that are able to reach around the cranium. Sliding calipers are used when chord measurements are taken that do not require the bow-like arms of the spreading calipers – as seen in Figure 23 above.

Most anthropologists generate an Osteometric "short form" that lists each measurement. This form may conveniently mirror the order of the variables

Bare Bones: A Survey of Forensic Anthropology

used in the Fordisc discriminant function computer program, making input of each measurement more convenient and accurate. Table 3 shows an example of a typical measurement form using selected variables.

FORDISC SHORT FORM - CRANIAL MEASUREMENTS

TABLE 3: *Cranial measurements used in the discriminant function program Fordisc. (Moore-Jansen et al., 1994).*

	LEFT	RIGHT			LEFT	RIGHT
1. Maximum length (*g-op*)	____		13. Nasal height (*n-ns*)	____		
2. Maximum breadth (*eu-eu*)	____		14. Nasal breadth (*al-al*)	____		
3. Bizygomatic breadth (*zy-zy*)	____		15. Orbital breadth (*d-ec*)	____	____	
4. Cranial height (*ba-b*)	____		16. Orbital height (*OBH*)	____	____	
5. Cranial base length (*ba-n*)	____		17. Biorbital breadth (*ec-ec*)	____		
6. Basion-prosthion length (*b-pr*)	____		18. Interorbital breadth (*d-d*)	____		
7. Maximum alveolar breadth (*ecm-ecm*)	____		19. Frontal chord (*n-b*)	____		
8. Maximum alveolar length (*pr-alv*)	____		20. Parietal chord (*b-l*)	____		
9. Biauricular breadth (*au-au*)	____		21. Occipital chord (*l-o*)	____		
10. Upper facial height (*n-pr*)	____		22. Foramen magnum length (*ba-o*)	____		
11. Minimum frontal breadth (*ft-ft*)	____		23. Foramen magnum breadth (*FOB*)	____		
12. Upper facial breadth (*fmt-fmt*)	____		24. Mastoid length (*MDH*)	____	____	

FORDISC SHORT FORM – MANDIBULAR MEASUREMENTS

TABLE 4: *Mandibular measurements used in the discriminant function program Fordisc. An asterisk denotes measurements that require a mandibulometer (Moore-Jansen et al., 1994).*

	LEFT	RIGHT		LEFT	RIGHT
25. Chin height (*gn-id*)	____		30. Minimum ramus breadth	____	____
26. Body height at mental foramen	____	____	31. Maximum ramus breadth	____	____
27. Body thickness at mental foramen	____	____	32. Maximum ramus height *	____	
28. Bigonial diameter (*go-go*)	____		33. Mandibular length *	____	
29. Bicondylar breadth (*cdl-cdl*)	____		34. Mandibular angle *	____	

FORDISC SHORT FORM – SELECTED POST-CRANIAL MEASUREMENTS

TABLE 5: *Selected postcranial measurements used in the discriminant function program Fordisc. (Moore-Jansen et al., 1994).*

HUMERUS	LEFT	RIGHT	FEMUR	LEFT	RIGHT
40. Maximum length	____	____	60. Maximum length	____	____
41. Epicondylar breadth	____	____	61. Bicondylar length	____	____
42. Max. vertical diameter of head	____	____	62. Epicondylar breadth	____	____
43. Max. diameter at midshaft	____	____	63. Maximum diameter of head	____	____
44. Min. diameter at midshaft	____	____	64. A-P subtrochanteric diameter	____	____

RADIUS	LEFT	RIGHT	TIBIA	LEFT	RIGHT
45. Maximum length	____	____	69. Condylo-malleolar length	____	____
46. Sagittal diameter at midshaft	____	____	70. Max. prox. epiphyseal dia.	____	____
47. Transverse diameter at midshaft	____	____	71. Max. dist. epiphyseal dia.	____	____

ULNA	LEFT	RIGHT	FIBULA	LEFT	RIGHT
48. Maximum length	____	____	75. Maximum length	____	____
49. Dorso-volar diameter	____	____	76. Max. diameter at midshaft	____	____
50. Transverse diameter	____	____			

The use of Osteometric points will be addressed in each chapter and metric analyses to determine sex, stature, and ancestry are covered.

Now that the reader has a basic understanding of human osteology, let's read forward and learn about how the knowledge of the human skeleton can contribute to skeletal identification.

Bare Bones: A Survey of Forensic Anthropology

References

White TD and Folkens PA (2005) *The Human Bone Manual*. London: Elsevier Academic Press.

Matshes E, Burbridge B, Sher B, Mohamed A, Juurlink B (2005) *Human Osteology & Skeletal Radiology: An Atlas and Guide*. Boca Raton, Fla.: CRC Press.

Buikstra JE, Ubelaker DH, editors (1994) *Standards for Data Collection from Human Skeletal Remains*. Arkansas Archeological survey Research Series No 44.

Moore-Jansen PM, Ousley SD, Jantz RL (1994) *Data Collection Procedures for Forensic Skeletal Material*. Report of Investigations No 48. The University of Tennessee, Knoxville, Department of Anthropology.

Sample Test Questions

1. The total number of bones in a human skull (including the hyoid) is:

    a. 8

    b. 16

    c. 29

    d. 44

2. A sacralized 5th lumbar vertebra is:

    a. Known as a "transitional" vertebra

    b. Is an example of idiosyncratic variation?

    c. Is helpful for radiographic comparison

    d. All of the above

3. An osteometric point is:

    a. A defined point or area of the skeleton from which measurements are made

    b. The spot on the cranium that corresponds to the hairline in a mature male

    c. A theory of skeletal functional morphology

    d. A type of caliper used to make measurements on the skull

Section II

Determination of Biological Sex

Sexual dimorphism is the difference in body form—both size and shape—between males and females of the same species. The degree of sexual dimorphism varies greatly between different species of mammals; in some species, the males and females are relatively the same size, whereas in other species, the females are actually larger than the males. In most mammals, the males are larger than the females. Humans exhibit an average 15 percent difference in size between males and females.

Several theories have been offered to explain sexual dimorphism. Charles Darwin was the first to hypothesize that disparity in size and certain traits within a species are the result of the male's attempts to attract mates. This phenomenon is considered to be a specific type of natural selection called **sexual selection**. This theory suggests that males have evolved an overall larger body size in order to win competitions against other males for copulatory rights with a particular female. Additionally, a larger male may represent a healthy, more fit parent better suited to protect the female and their young from predators. When the males and females of a species differ significantly in size or shape, they are better able to utilize resources by eliminating competition for the same food—a phenomenon known as niche divergence. Finally, males and females may have different optimum body sizes that protect each individual from environmental and nutritional stress. When considering the overall size of the female, one must also calculate the energetic costs of gestation and lactation. Large females may be at a disadvantage during times of stress when carrying or providing for offspring, relative to smaller females.

Whether a single theory or multiple theories apply, in humans, males are larger and more robust than females. This is readily observable in the skeleton and provides good evidence for the biological sex of the decedent.

DETERMINATION OF BIOLOGICAL SEX

The determination of sex is usually one of the first aspects of the biological profile that is considered. Identifying an unknown skeleton as either male or female will automatically rule out 50 percent of the population. Additionally, the determination of many other aspects of the biological profile is sex-specific, thus requiring an a priori knowledge of sex (France 1998). Thus, the determination of sex is always a good starting point.

Biological sex is difficult to determine in prepubescent children. Indicators of sex in the adult skeleton are related to the secondary sexual characteristics that develop during puberty. Attempts to identify prepubescent traits have enjoyed limited success. Boucher (1957) devised a method for evaluating the depth and width of the sciatic notch that enabled her to correctly estimate sex between 57.8 percent and 95.1 percent of the time, varying with ancestry. However, later researchers have failed to replicate her results (Weaver 1980). A simple assessment of the elevation of the auricular surface of the sacroiliac joint has been found to be the most promising technique; however, most anthropologists find sex determination of juvenile remains extremely risky, if not impossible (Weaver 1980; Mittler and Sheridan 1992).

This chapter will be restricted to determination of sex in adults.

Primary sex characteristics develop during gestation of the embryo and are present at birth. These include sex chromosomes, external genitalia, testes, and ovaries. However, none of these can be detected from overall gross examination of the skeleton. During the time of puberty, humans begin to develop

Bare Bones: A Survey of Forensic Anthropology

secondary sex characteristics as a result of increased estrogen or testosterone. Many of these newly acquired traits are quite apparent in the skeleton. Males begin to grow taller, get a deeper voice, and grow facial hair. On the other hand, women begin to develop breasts, start to menstruate, start to develop more subcutaneous body fat, and grow wider hips. Less obvious are the secondary sex characteristics that begin to occur in the skeleton. One of the most obvious skeletal changes that can be noted in males and females is the overall size of the bones. As a result of sexual dimorphism, male bones tend to be overall larger and more **robust** than females, who have smaller and more **gracile** bone structures. However, it is important to remember that humans exhibit a great deal of natural variation, and it is possible to have small, gracile males and larger, robust females.

Sex determination is performed using two different methods: **nonmetric** and **metric** assessment. Nonmetric methods for determining sex involve a visual analysis of the bones for various morphological characteristics. These traits are ones that are not easily measured and instead vary by either their mere presence or absence or on a discontinuous scale such as small, medium, or large. On the other hand, metric analysis involves taking a battery of standardized measurements of the skeleton and applying various statistical techniques to analyze those data.

NONMETRIC ASSESSMENT OF SEX

At first glance at the human skeleton, it can be seen that there is an obvious size and shape difference between male and female bones. Visual analysis shows that males tend to have larger and longer bones with more pronounced markings for muscle attachments. Similarly, joint surfaces in males are larger than in females. Females on the other hand tend to have smaller and shorter bones and smaller sites for muscle attachments. While the size and shape differences between male and female skeletons are quite apparent, there are two specific areas in the skeleton that will lead to a more diagnostic assessment of sex: the pelvis and the skull.

Pelvic Morphology

The pelvis has important implications for locomotion and childbirth. As a result of our unique bipedal form of locomotion, the human pelvis has adapted to allow for an efficient gait. Equally important are the structural demands placed on the pelvis due to childbirth. Female pelvic configuration must allow for the passage of the head of the human fetus—which is particularly large relative to other mammals—through the birth canal. The gait of females is slightly less efficient than that of males, since the hips are relatively wider. This results in a greater angle at the knees in females than males. This angle, known as a "q" angle, creates more lateral instability in females. The readers will have noticed that among world-class female sprinters, the most successful athletes have relatively narrow hips.

Therefore, pelvic morphology in females is a compromise between efficient locomotion and childbirth. As such, there are several differences between the male and female pelves that make it the best area in the skeleton for determining sex (**Table 1**: Male and female pelves). On the whole, females have wider pelves with a large, circular birth canal. Male pelves are narrower with a smaller and more triangular pelvic outlet. As was stated in Chapter 4, the pelvis is composed of three different bones: the pubis bone, the ischium, and the ilium. Of these three bones, the pubis is particularly helpful in differentiating between male and female pelves.

TABLE 1: *Sex differences in pelvic morphology*

TRAITS	MALE	FEMALE
Pelvis overall	Massive, rugged, marked muscle attachment sites	Less massive, gracile, smoother
Symphysis	Higher	Lower
Subpubic angle	Acute, v-shaped	Obtuse, u-shaped
Obturator foramen	Large, often ovoid	Small, triangular
Acetabulum	Large, tends to be directed laterally anterolaterally	Small, tends to be directed
Greater sciatic notch	Asymmetrical, deep, and narrow	Symmetrical, shallow, and wider
Sacroiliac joint	Large	Small and oblique
Preauricular sulcus	Less frequent	More frequent, better developed
Ilium	High, tends to be vertical	Lower, laterally divergent
Sacrum	Longer, narrower	Shorter, broader
Pelvic brim, or inlet	Heart-shaped	Circular, elliptical
Pelvic outlet	Relatively smaller	Oblique, shallow, spacious

SOURCE: *Krogman and İşcan 1986, 209.*

Bare Bones: A Survey of Forensic Anthropology

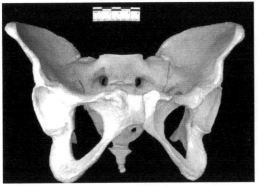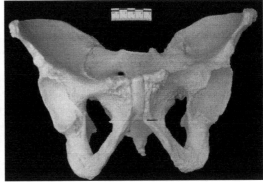

■FIGURE 1A AND 1B: *Anterior views of a female pelvis (left) and male pelvis (right).*

There are three traits of the pubis that are reported to sex the skeleton with 95 percent accuracy (Phenice 1969). These traits are the **ventral arc**, the subpubic concavity, and the ischiopubic ramus. The ventral arc is an elevated ridge of bone located on the ventral aspect of the female pubis. It is found primarily in females; however, it is important to not confuse the ventral arc with a more medially positioned crest found in males (Phenice 1969; Anderson 1990). Use of this trait has been found to sex the skeleton with 96 percent accuracy (Sutherland and Suchey 1991). The area immediately inferior to the pubic bone is known as the **subpubic concavity**. This area is more concave in females and more convex in males (Phenice 1969). When looking at the right and left pubic bones in anatomical position, the **subpubic angle** is u-shaped in females and v-shaped in males (Krogman and İşcan 1986). The **ischiopubic ramus** is a portion of bone that connects the pubis and the ischium. When looking at the medial aspect of this area, it is narrow in females with a slight crest along the midline and broad and flat in males (Phenice 1969). Additionally, the female pubis is longer and wider than the male pubis, creating a more rectangular pubic bone in females and a more triangular pubis in males (see photos of pelves in Figures 1A and 1B).

Several other areas of the pelvis are helpful for sexing the skeleton. The **auricular surface** is the ear-shaped area on the ilium that serves as the junction between the ilium and the sacrum. In females, the auricular surface is elevated, whereas in males it is flatter. The **preauricular surface** is a depression located between the sacroiliac junction and the sciatic notch. This trait is characteristically found in females and is usually absent in males. The **sciatic notch**, located on the posterior aspect of the pelvis inferior to the posterior inferior iliac spine, is wider in females and narrower in males (see **table 1**).

Cranial Morphology

The cranium is the second best place to look when determining sex from the skeleton. Overall, the cranium is larger and more rugged in males and smaller and more gracile in females. Looking at the areas of muscle attachments can often prove helpful, as males tend to have more pronounced areas of muscle insertions, whereas female muscle-attachment points are smaller. The area just above the eye orbits is known as the superciliary arches, supraorbital torus—or more simply put, the **brow ridge**. This area is more pronounced in males and smaller in females. The forehead, or frontal bone, tends to slope posteriorly in males, while in females it is more vertical. The upper margin of the eye orbit is most often sharp in females and more rounded in males. The posterior extension of the zygomatic arch can sometimes take the form of a ridge or a crest just above the external auditory meatus (the opening for the ear canal). This ridge, or **supramastoid crest**, is most common in males. The most posterior portion of the occipital bone is the location of attachment for the nuchal, or neck, muscles. This area is more pronounced and larger in males and in some cases can even appear in the form of a beaklike **inion hook**. The **mastoid process** is a bony protrusion that is located inferior and slightly posterior to the external auditory meatus. This protrusion is larger in males and smaller in females.

Bare Bones: A Survey of Forensic Anthropology

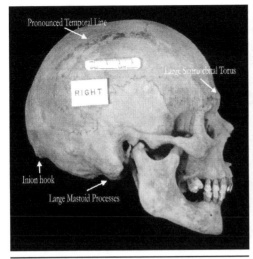

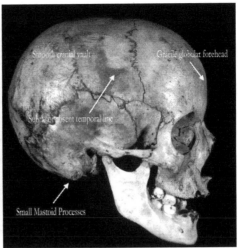

■Figure 2a: *Male cranium (lateral view).*   ■Figure 2b: *Female cranium (lateral view).*

The mandible can also be used to assess the sex of a skeleton. The mandible is similar to the rest of the skeleton in that it is larger and more robust in males and smaller and more gracile in females. The chin, or **mental eminence**, tends to be square in males and rounded or pointed in females. **The mandibular angle**, located at the posterior portion of the mandible, is usually less than a 120-degree angle in males and more obtuse in females. The junction between the body of the mandible and the ramus is known as the gonial region (Parr 2005, 2006).

TABLE 2: *Nonmetric indicators of sex in the cranium*

CRANIAL MORPHOLOGY

MALES	FEMALES
Large rugged cranium	Smaller gracile cranium
Large brow ridges	Small or no brow ridges
Dull superior orbital margins	Sharp superior orbital margins
Receding forehead	Vertical globular forehead
Large mastoid processes	Small mastoid processes
Square chin	Rounded chin
Marked temporal line	Gracile nuchal area

Postcranial Morphology

The postcranial skeleton in sex determination is most helpful in terms of its size and shape. Male postcranial bones are usually larger and longer than female bones and have larger areas for muscle attachments. The femur, in particular, is longer in males and has a larger **femoral head**, whereas females have a shorter and smaller femoral head (see metric analysis below). Additionally, the **linea aspera**, a raised line located on the posterior aspect of the femur and an insertion area for several muscles of the thigh, is much more pronounced and elevated in males than in females.

The **costal cartilage** that lies between the end of the rib and the sternum has different ossification patterns in males and females. In females, the costal cartilage ossifies more centrally, whereas in males, the ossification occurs along the margins of the ribs.

METRIC ASSESSMENT OF SEX

One of the first steps when performing a skeletal analysis is to take standardized measurements of the skull and the postcranial skeleton. These measurements can then be used in various statistical formulae to help identify the sex of an individual. Additionally, as an objective measure, the metric technique of sexing the skeleton is a useful way to check the morphological assessment of sex (Stewart 1979).

Discriminant function analysis allows the researcher to predict group affiliation based on metric measurements. Typically, discriminant function analysis uses continuous measurements to separate individuals into two or more different categories. This approach can be used for sexing the skeleton.

Cranial Metric Assessment

Giles and Elliot (1963) analyzed 408 European American and African American crania from the Terry and the Hamann-Todd collections by taking nine standardized measurements of the cranium and applying those measurements to discriminant function analysis. They determined that sex could be accurately determined from the cranium 82–89 percent of the time.

While the Giles and Elliot method has proven quite accurate, it can only be performed when the cranium has retained the necessary osteometric landmarks. Thus, Holland (1986) developed a method for sex determination from the base of the cranium. Holland took nine measurements on the cranial base of a hundred individuals, equally divided by sex and ancestry, from the Terry collection. From these measurements, six regression equations were created from which sex was accurately predicted with 71–90 percent accuracy.

TABLE 3: *Discriminant function analysis using cranial measurements*

DISCRIMINANT FUNCTION ANALYSIS FOR WHITES AND BLACKS USING CRANIAL MEASUREMENTS

VARIABLE	BLACKS	WHITES
Maximum Breadth	–0.23287	–0.10646
Bizygomatic Breadth	0.40358	0.40376
Basion-Bregma	0.15754	----
Nasion-Prosthion	0.20526	----
Nasal Height	----	0.37828
Orbit Height	----	–0.29159
Frontal Chord	----	0.19283
Parietal Chord	0.26434	0.08209
Constant	–84.6585	–76.29420

SOURCE: *Jantz and Moore-Jansen 1998.*

The development of the Forensic Data Bank (FDB) at the University of Tennessee provided practicing anthropologists with data derived from a contemporary population with which to compare data from unknown specimens. The Fordisc computer program, utilizing data from the Forensic Data Bank, not only uses the discriminant function statistical method first applied by Giles and Elliot, but also enables the examiner to choose from a variety of variables and population groups to best determine the sex and ancestry of an unknown skeleton (Jantz and Ousley [1993–1996] 2005). The Fordisc program, now in its third version, has become the standard among forensic anthropologists for metric analysis. The statistical output specifically addresses issues of probability and error that have become crucial to rules of evidence and admissibility of scientific evidence.

Table 4 shows the output for a typical Fordisc assessment of biological sex. Note that the assessment provides a classification (in this case "male"), a statement of posterior probability that the assessment is correct, and the typicality of the specimen relative to the group means for the variables, as well as a table listing the measurements of the unknown specimen, the means of each group for each variable, and the relative weight assigned to each variable based on that variable's ability to discriminate between the two groups.

TABLE 4: *Output from the Fordisc discriminant function computer program*

FORDISC REFERENCE GROUP CLASSIFICATION USING
21 VARIABLES

FROM GROUP	INTO GROUP M	F	TOTAL COUNTS	PERCENT CORRECT
M	176	26	202	87.1 %
F	19	144	163	88.3 %
Totals:	320		365	87.7 %

TWO GROUP DISCRIMINANT FUNCTION RESULTS

GROUP	CLASSIFIED INTO	DISTANCE FROM	POSTERIOR PROBABILITIE	TYPICALITY
M	** M **	27.6	.999	.153
F		42.5	.001	.004

TWO GROUP DISCRIMINANT FUNCTION COEFFICIENTS

		M 202	F 163	D.F. WEIGHT	RELATIVE WEIGHTS
GOL	179	186.60	177.82	.075	5.9 %
XCB	145	139.25	134.40	.043	1.9 %
ZYB	141	130.79	121.72	.486	39.3 %
BBH	137	137.69	130.05	.030	2.0 %
BNL	103	103.92	97.20	.154	9.3 %
BPL	94	99.46	94.63	−.038	1.6 %
MAB	67	63.75	60.37	.009	0.3 %
MAL	50	55.46	52.75	−.012	0.3 %
AUB	124	122.32	115.65	−.200	11.9 %
WFB	99	97.29	93.40	−.110	3.8 %
NLH	57	52.01	48.52	.159	4.9 %
NLB	25	24.79	23.49	−.094	1.1 %
OBB	41	40.36	38.17	.020	0.4 %
OBH	37	33.85	33.66	−.295	0.5 %
EKB	102	99.05	94.44	.034	1.4 %
DKB	26	22.94	21.71	.009	0.1 %
FRC	105	113.07	107.40	.052	2.6 %
PAC	116	117.96	114.19	.012	0.4 %
OCC	97	97.81	94.46	−.056	1.7 %
FOL	35	36.96	35.09	.021	0.4 %
MDH	35	32.15	27.71	.262	10.4 %
Constant		−66.403			

In this case, metric analysis strongly indicates that the unknown decedent is a male. However, the low "typicality" shows that the measurements vary significantly from the means for the male group.

Postcranial Assessment

In the postcranial skeleton, the size of the articular surfaces can be used to accurately sex the skeleton. The femoral head is typically larger in males and smaller in females. In males, the femoral-head diameter is usually larger than 46.5 mm., whereas in females, the femoral-head diameter is smaller than 43.5 mm. Measurement between 43.5 and 46.5 mm. represent the overlap between the two groups and so would remain undetermined (Stewart 1979).

FEMORAL-HEAD DIAMETER		
Diameter > 47.5	mm =	Male
Diameter 46.5 – 47.5	mm =	Probably Male
Diameter 46.5 – 43.5	mm =	Undetermined
Diameter 42.5 – 43.5	mm =	Probably Female
Diameter < 42.5	mm =	Female

Similarly, the humeral head is also typically larger in males than females. A humeral-head diameter of greater than 47 mm. strongly suggests a male, and a diameter of less than 43 mm. would almost always represent a female (Stewart 1979).

HUMERAL-HEAD DIAMETER		
Diameter > 47	mm =	Male
Diameter 46 – 47	mm =	Probably Male
Diameter 46 – 44	mm =	Undetermined
Diameter 43 – 44	mm =	Probably Female
Diameter < 43	mm =	Female

Jantz and Moore-Jansen (1998) have developed a discriminant function method for determining sex from the femur using four measurements: the maximum length of the femur, bicondylar breadth, epicondylar breadth, and the maximum diameter of the head. This method has an 89 to 98 percent accuracy rate.

Other Aspects of Sex Determination

Parturition Scars

There are two areas on the pelvis that have been associated with childbirth: the **preauricular sulcus** and **dorsal pitting** of the pubis. These pits, or sulci,

Bare Bones: A Survey of Forensic Anthropology

are locations of ligamentous attachments (Kelley 1979). The preauricular sulcus is located on the ilium, just inferior and anterior to the auricular surface. It serves as the attachment point for the anterior sacroiliac ligament. Dorsal pits are found on the dorsal aspect of the medial pubis.

Considerable debate has ensued over these two areas as to whether or not they are truly an artifact of childbirth. Suchey and associates (1979) conducted a study on 486 American females to see the prevalence of dorsal pits. They found that there was a weak statistical correlation between females who underwent full-term pregnancies and dorsal pits. Additionally, they found that some women who had never been pregnant still displayed significantly large pits. Similar results have been found with the preauricular sulcus. Spring and colleagues (1989) noted that while the preauricular sulcus is sometimes present in women who have gone through childbirth, it is also found in nulliparous females. The mechanism of the pitting has also been thoroughly debated, with theories ranging from birth trauma to avulsion and resorption of bone related to the hormone relaxin, which softens the joints of the pelvis during pregnancy.

No matter the relationship between dorsal pitting and preauricular sulci to parturition, evidence of either is a strong indicator that the remains are those of a female.

SUMMARY

Humans are sexually dimorphic, with males being—as a group—significantly larger than females. The robusticity in males, in combination with the unique anatomical constraints placed on females related to childbirth, permit anthropologists to determine biological sex from the skeleton. The biological sex of a decedent can be determined from analysis of the skeleton in over 90 percent of cases, especially when the pelvis is present for examination. Once the sex has been determined, the anthropologist can proceed to determining age at death, ancestry, and stature—completing the biological, or demographic, profile.

References

Anderson, B. E. 1990. Ventral arc of the os pubis: Anatomical and developmental considerations. *American Journal of Physical Anthropology* 83:449–458.

Boucher, B. J. 1957. Sex differences in the foetal pelvis. *American Journal of Physical Anthropology* 15 (4): 581–600.

France, D. L. 1998. Observational and metric analysis of sex in the skeleton. In *Forensic osteology: Advances in the identification of human remains*, ed. K. J. Reichs, 163–186. Springfield, IL: Charles C. Thomas.

Giles, E., and O. Elliot. 1963. Sex determination by discriminant function analysis of crania. *American Journal of Physical Anthropology* 21:53–68.

Holland, T. D. 1986. Sex determination of fragmentary crania by analysis of the cranial base. *American Journal of Physical Anthropology* 70:203–208.

Jantz, R. L., and P. H. Moore-Jansen. 1998. A database for forensic anthropology: Structure, content, and analysis. Report of Investigations No. 47, Department of Anthropology, Univ. of Tennessee at Knoxville.

Jantz, R. L., and S. D. Ousley. [1993-1996] 2005. FORDISC: Personal computer forensic discriminant functions. 3rd vers. Knoxville: University of Tennessee.

Kelley, M. A. 1979. Parturition and pelvic changes. *American Journal of Physical Anthropology* 51 (4): 541–546.

Krogman, W. M., and M. Y. İşcan. 1986. *The human skeleton in forensic medicine*. Springfield, IL: Charles C. Thomas.

Parr, N. M. 2006. An assessment of nonmetric traits of the mandible used in the determination of ancestry. *Proceedings of the American Academy of Forensic Sciences* 12:306.

————. 2005. Determination of ancestry from discrete traits of the mandible. Master's thesis, Univ. of Indianapolis.

Pearson, G. A. 1996. Of sex and gender. *Science* 274:328–329.

Phenice, T. W. 1969. A newly developed visual method of sexing the os pubis. *American Journal of Physical Anthropology* 30:297–301.

Spring, D. B., C. O. Lovejoy, G. N. Bender, and M. Duerr. 1989. The radiographic preauricular groove: Its nonrelationship to past parity. *American Journal of Physical Anthropology* 79:247–252.

Stewart, T. D. 1979. *Essentials of forensic anthropology*. Springfield, IL: Charles C. Thomas.

Suchey, J. M., D. V. Wiseley, R. F. Green, and T. T. Noguchi. 1979. Analysis of dorsal pitting in the os pubis in an extensive sample of modern American females. *American Journal of Physical Anthropology* 51:517–540.

Sutherland, L. D., and J. M. Suchey. 1991. Use of the ventral arc in pubic sex determination. *Journal of Forensic Sciences* 36:501–511.

Washburn, S. L. 1948. Sex differences in the pubic bone. *American Journal of Physical Anthropology* 6:199–207.

Weaver, D. S. 1980. Sex differences in the ilia of a known sex and age sample of fetal and infant skeletons. *American Journal of Physical Anthropology* 52:191–195.

Sample Test Questions

1. The best bone to use in determining biological sex is:

   a. Cranium

   b. Pelvis

   c. Long bones

   d. Midface

2. If you only have the skull available to you for analysis, which traits should you use to determine sex?

   a. Size of the cranium

   b. Height of the frontal bone

   c. Size and shape of the mandible

   d. A and C

3. The difference in size and shape between males and females is called:

   a. Sexual selection

   b. Growth potential

   c. Sexual dimorphism

   d. Delayed neoteny

4. In discriminant function analysis, the sectioning point is best described as:

   a. The specific osteometric point used

   b. The point between 2 group means that best discriminates members of each group with least error

   c. The upper limit at which a person may be male

   d. The coefficient that is applied to each variable

5. **The shape of the female pelvis is an evolutionary compromise between what 2 factors?**

    a. Locomotion and maximum load potential

    b. Locomotion and childbirth

    c. Childbirth and uterine pathology

    d. A and C are correct

6. **Why can an anthropologist easily determine the sex of a skeleton but physicians have difficulty?**

    a. There is no clinical need for physicians to determine the sex of a skeleton

    b. Anthropologists are more highly trained

    c. Anthropologists learn more anatomy in school than physicians

    d. All of the above

7. **If a skeleton exhibits characteristics of both males and females**

    a. Then the decedent must have been of mixed ancestry

    b. The decedent may have been a transsexual

    c. No absolute statement can be made concerning the sex of this decedent

    d. It is impossible for a skeleton to exhibit both male and female characteristics

8. **An explanatory theory for sexual dimorphism that focuses on the use of different ecological resources by males and females of a species is:**

    a. Sexual selection

    b. Male's role as protector of the group

    c. General selection for large body size

    d. Niche divergence

Determination of Ancestry

Of all the aspects of the biological profile, the determination of ancestry is the most difficult to determine and the most controversial due to the complexity of traits and the problems in understanding the differences between the biological and social aspects of ancestry (Reichs 1986). The racial categories used in Western societies were constructed more by historical accident than by biological reality. Racial categories offer little basis for understanding the full extent of human biological variation. Most biological anthropologists and forensic anthropologists would agree that racial boundaries are dynamic in nature and thus are more appropriately viewed as a series of gradually changing populations that have similar characteristics based on their geographical location. The features that we use to classify races—skin color, hair texture, nasal width, and others—are clinally distributed, and these clines differ for each character trait.

THE BIOLOGICAL CONCEPT OF RACE

The biological concept of "race," as defined by Ashley Montagu (1962), is the existence of "different populations of the same species, which are distinguished from one another by the possession of certain distinctive hereditary traits." Anthropologists have used the race concept for many years as a way to group populations in order to do comparative studies on human variation. For years, scientists have been divided into two groups: those who wish to separate individuals in to biologically distinct races and those who feel that racial boundaries are plastic and the human population cannot be separated into different racial categories. The eighteenth-century botanist Carolus Linnaeus was the first person to divide groups of people into human varieties (later called "races"), based on his assumption of the fixity of species. He separated the human species into four subspecies: *H. americanus*, the red-colored people that inhabited the Americas; *H. europus*, the white-colored people who inhabited Europe; *H. asiaticus*, the yellow-colored people who inhabited Asia; and *H. afer*, the black-colored people who inhabited Africa.

Johann Blumenbach was a student of Linnaeus and expanded on his classification scheme by including a fifth race and changing the names of the four original races. Thus, Blumenbach's five races were called American, Caucasoid, Mongoloid, Ethiopian, and the new group, the Malay. Blumenbach believed that the Caucasoids were direct descendants of Adam and Eve and all other races had *devolved* from the Caucasoids. Consequently, Blumenbach's classification scheme is hierarchical and is a radical departure from Linnaeus's primarily geographical classification.

Both Linnaeus and Blumenbach may appear racist to the modern observer; however, in reality, they were quite progressive for their time. While many of the scientists of the 1700s and 1800s believed that the major human populations could be divided into different races, they did not adhere to the notion that races were static with sharp boundaries between them. Instead, they thought that each race could change due to environmental circumstances. Nonetheless, Linnaeus's and Blumenbach's classification schemes have been used by many individuals to propagate the superiority of one race over another, and Blumenbach's system exists to this day and is often used to disseminate racist ideologies.

PROBLEMS WITH THE "RACE" CONCEPT

In the mid-1800s, Charles Darwin introduced his theories of evolution and natural selection to the scientific community. Slightly later, Gregor Mendel discovered discrete units (now known as genes), which are passed to offspring, allowing favorable traits to be passed on to future generations. With the combination of Darwin's ideas of natural selection and Mendel's discovery of genes, perspectives of race began to change. Previously, race was considered

Bare Bones: A Survey of Forensic Anthropology

relatively malleable; however, in the early 1900s, each race became static and unchangeable, where people were born into particular races, each of which had a suite of specific, innate attributes. This concept is known as **biological determinism** and states that individuals are born with a specific set of genes that determines their behavioral norms.

The eugenics movement stemmed from the concept of biological determinism. Proponents of eugenics were concerned with the improvement of society through the regulation of reproduction, family size, and fertility, according to the genes of the parents. This movement promoted a biological basis for racism. Francis Galton was one of the foremost advocates of eugenics. He believed that anything observed in an individual, such as physical traits or behavior, was the result of genes. Therefore, the characteristics of different races are inherent and due to biology. According to the eugenics movement, the races are static and unchangeable as proven by biology. The eugenics movement not only permeated the scientific realm but also was disseminated throughout all aspects of American life. Institutions were built to house the "feebleminded," and state and federal laws were passed to perform sterilization on those deemed "unfit" to reproduce.

CURRENT IDEAS ON "RACE"

In the 1940s, an increased knowledge of genetics and human variation caused scientists to change their views on race. Scientists began to realize that differentiating the human population into races had severe consequences, such as the justification of slavery, discrimination, and genocide (Littlefield et al. 1982). Soon thereafter, anthropologists began to doubt whether the different races were truly valid and began seeing race as merely a gradation of human variation (Lieberman and Jackson 1995). Research done as part of the Human Genome Project has shown that there is more variation within a group than is seen between groups.

As the years have passed, there has been a steady decline in the number of scientists who believe that race exists on a biological level (Littlefield et al. 1982). It is now commonly believed that racial categories are not biological but instead represent an individual's social or ethnic affiliation (Rathbun and Buikstra 1984). In fact, it has even been suggested that the term "race" be abandoned completely and replaced by another term that does not carry negative social connotations (Lieberman and Jackson 1995). In forensic anthropology, the term **ancestry** is commonly used instead of "race" to refer to a group of individuals that are descended from the same ancestral population.

IF RACE DOESN'T EXIST, WHY IS IT STILL USED IN FORENSIC ANTHROPOLOGY?

If most anthropological communities have abolished the concept of race, why do forensic anthropologists continue to address ancestry in their studies of human skeletal remains? In order to identify an individual, a biological profile must be constructed that includes age, sex, and ancestry. While most scientists abstain from placing individuals in separate ancestral groups, the common layperson usually has a self-prescribed notion of his or her "race." Forensic anthropologists will continue to classify individuals into separate categories that are equivalent to discrete races as long as society as a whole does the same (Gill 1998). Thus, the custom of assigning individuals to ancestral categories is a practical one in forensic framework. Consequently, forensic anthropologists continue to include ancestry determination in their studies of human variation as a means to assist in the identification of a decedent.

Many anthropologists believe that forensic anthropologists reinforce social and racial stereotypes by classifying individuals by ancestral affiliation. Forensic anthropologists are well aware of this problem and the dilemma that it causes. But forensic anthropologists are applied scientists who work within the context of social history. Until society no longer considers ancestry part of an individual's personal identity, anthropologists are bound to continue to assess skeletal remains for clues as to the decedent's ancestral heritage.

DETERMINATION OF ANCESTRY

Despite the controversy surrounding the race issue, anthropologists are still able to differentiate between broadly dispersed geographic groups, especially populations living in areas with geographic barriers to gene flow (e.g., oceans or deserts). Even anthropologists with theoretical or methodological issues regarding determination of ancestry agree that it is often possible to differentiate the most broadly defined populations, such as European, African, and

Bare Bones: A Survey of Forensic Anthropology

Native American. Over the years, much research has gone into determining ancestry from the human skeleton. Two methods are frequently used to assess ancestral affiliation. The nonmetric method is a visual method that involves looking at morphological characters of the skull and scoring them on a discrete or discontinuous scale. The metric technique takes the standardized measurements of the skull and applies a battery of statistical tests, such as discriminant function analysis, to those measurements (Parr 2005).

NONMETRIC ANALYSIS

Nonmetric traits are normal skeletal variants that are visually determined and not pathological or traumatic in nature. These traits vary between males and females and between different populations and cannot be easily measured. Some individuals feel that nonmetric traits are not precise because they are not continuous and have high degrees of interobserver error. However, nonmetric traits are often used, because the analysis is simple to perform and does not require expensive or difficult-to-handle equipment. Additionally, the increase of the use of nonmetric traits has called for an increase of standardization of traits with precise descriptions and line drawings of pictures. As such, nonmetric analysis is becoming increasingly more accurate and is being used in studies all over the world.

Scoring of nonmetric traits is typically performed in two different ways. Truly **discrete** or **discontinuous traits** are scored as being "present or absent" or "complete or incomplete." These traits tend to be the presence/absence of a particular bone or foramen, or a foramen that has not fully closed. Traits with more intermediate stages are scored on a ranked scale, such as "0, 1, 2, 3." Some traits are considered **quasicontinuous** in that they appear to vary in a gradating order but are still difficult to measure in a continuous manner. Such traits are scored as "small, medium, or large."

The most frequently cited literature on ancestry determination is Stan Rhine's chapter in Gill and Rhine's book *Skeletal Attribution of Race* (1990). While this was a landmark study in its time, Rhine used small sample sizes and rudimentary statistics from which he made assumptions about the distribution of morphological traits between populations. Additionally, such a study reinforces the typological views held by anthropologists at the turn of the century. More recently, researchers have been working to improve the reliability of determining ancestry using nonmetric traits by increasing their sample sizes and using more detailed statistical tests.

No trait can be restricted exclusively to one population. Instead, traits vary from one population to the next, with certain traits occurring at higher frequencies in some populations. Two recent studies looked at the preponderance of morphological characteristics in various populations, and there are several traits that were found to be the most diagnostic between the ancestral groups. The **inferior nasal aperture** is the area at the most inferior part of the opening for the nose. This area may appear to have guttering, incipient guttering, a strait nasal sill, partial nasal sill, or a complete nasal sill. The contour of the **nasal bones** is the area across the bridge of the nose, which may be round, oval, triangular, or vaulted. The **nasal aperture** is the opening for the nose, which varies in width from narrow aperture, medium width, or wide aperture. Interorbital breadth is the distance between the eye orbits. This breadth may appear narrow, intermediate, or broad. **Nasal overgrowth** occurs when the inferior portion of the nasal bones project anteriorly past the maxillary bones and the superior portion of the nasal aperture. Overgrowth is either present, absent, or unobservable. The **transverse palatine sutures** are located on the inferior surface of the bony palate. These sutures vary in appearance as straight, symmetrical; anterior bulging, symmetrical; anterior and posterior bulging, scalaris; and posterior bulging, symmetrical. The **post-bregmatic depression** is a small concavity located on the sagittal suture just posterior to the osteometric point bregma. This trait is either present or absent. The **supranasal suture** is a complex, interlocking suture located just superior to the nasal bones and traverses the osteometric point glabella. This suture has various states such as open, closed but visible, closed, barely visible, or obliterated (Hefner 2003).

The mandible also has several diagnostic areas for ancestral determination. The degree of **ramus inversion** can be observed as a medial rotation of the posterior portion of the mandibular ramus. This area may have no inversion, slight inversion, moderate inversion, or extreme inversion. The gonial area is located at the inferior and posterior portion of the mandibular corpus at the junction with the mandibular ramus. This area may be analyzed for the amount of **gonial flare** present. Gonial flare may appear as inverted, no flare, slight flare, moderate flare, or extreme flare. There are various areas of muscle-attachment sites along the mandible. The most obvious are those located along the medial border of the ramus, created by the attachment of the medial pterygoid muscle. This area may appear gracile with little to no ridging, slight

Bare Bones: A Survey of Forensic Anthropology

ridging, moderate ridging, or robust with extreme ridging. Chin shape has been found to vary between different ancestral populations. The shape of the chin may have a round, square, or pointed appearance. The **mental foramen** is a small hole for blood vessels and nerves located on either side of the chin, typically near the cheek teeth. This foramen can vary in number and position. Usually, there is only one mental foramen on either side of the mandible; however, some individuals may have up to three foramina.

Native Americans

Individuals of Native American ancestry are most likely to possess the following character traits at a higher frequency than other populations: a straight inferior nasal aperture with incipient guttering, vaulted nasal bones, and a medium breadth nasal aperture. Also, narrow interorbital breadth and nasal overgrowth are more prevalent among Native Americans. The transverse palatine sutures are straight and symmetrical, while the postbregmatic depression is rarely present (Hefner 2003).

In many parts of the country, Native American remains are often prehistoric or historic era remains. These individuals often exhibit extreme tooth wear but retain all of their dentition without signs of carious lesions. The wear is related to grit in the diet, including abrasive particulates due to processing of corn with mortar and pestle. The absence of caries is due to the lack of processed sugar in the Native American diet (Figure 1: Tooth wear).

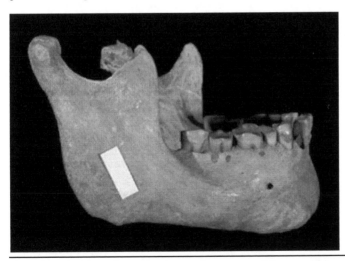

■FIGURE 1: *Note the extensive tooth wear on this mandible.*

East Asians

Straight or incipient guttering of the inferior aspect of the nasal aperture occurs at a higher frequency among East Asians than other geographic populations. The nasal bones are more likely to display a plateau shape: the nasal aperture is generally intermediate in width, while the interorbital breadth is most likely intermediate in width. Nasal overgrowth is relatively rare among East Asians, and the transverse palatine sutures are straight and symmetrical as in Native Americans (Hefner 2003).

African Ancestry

The inferior nasal aperture of individuals of African descent will present incipient guttering or full guttering at higher frequencies than other ancestral groups. Other nonmetric traits that occur at higher frequencies include a round, "quonset hut" nasal-bone contour, a wide nasal aperture, and broad interorbital breadth. Nasal overgrowth is seldom observed among individuals of primarily African ancestry. The transverse palatine sutures take on a symmetrical and anteriorly bulging appearance, and postbregmatic depression is commonly seen (Hefner 2003). Looking at the mandible, moderate to extreme ramus inversion is the norm, as well as inverted gonia. Individuals of African ancestry have also been noted to exhibit more robust muscle-attachment sites and a round chin. Furthermore, if the there are multiple mental foramina present, there is a greater likelihood that the individual is of African descent than another ancestral category. Additionally, the mental foramen is often placed in a more posterior position in Africans and African Americans than individuals of other ancestral groups (Parr 2005).

European Ancestry

Individuals of European ancestry typically have partial or full sill present on their inferior nasal aperture. The nasal-bone contour is most often triangular in shape, and the nasal aperture is usually narrow in width. Europeans have an intermediate interorbital breadth and sometimes display some nasal overgrowth. The transverse palatine suture bulges anteriorly and posteriorly at a higher frequency in this ancestral group. The supranasal suture may appear as an open suture, or it may be closed but still visible (Hefner 2003). The mandibular rami display little to no inversion, and the gonia are most frequently straight (neither inverted or everted). The European ancestral group often presents with more gracile muscle-attachment areas than the other groups, and individuals can present with either a round or square chin. There is usually only one mental foramen present, which is located more anteriorly on the mandible than in other ancestral groups (Parr 2005).

METRIC ANALYSIS

Determining ancestry using continuous measurements is similar to determining sex metrically. Standardized measurements are taken of the cranium and postcranial skeleton, which are then applied to various statistical formulae. One of the most common techniques is using discriminant function analysis, discussed in previous chapters.

Cranial Measurements

In 1962 Giles and Elliot examined 408 known individuals from the Hamann-Todd and Terry collections to see if standard cranial measurements could be used to assess ancestry. Using discriminant function, they determined that males and females could be placed into the appropriate ancestral categories 80 – 88% of the time (see Table 1 below).

TABLE 1: *Discriminant function weights for distinguishing American whites from American blacks on the basis of cranial measurements*

MEASUREMENT	MALE WEIGHTS	FEMALE WEIGHTS
Basion-prosthion	+3.06	+1.74
Glabello-occipital length	+1.60	+1.28
Max. width	-1.90	-1.18
Basion-bregma height.	-1.79	-0.14
Basion-nasion	-4.41	-2.34
Max. bizygomatic breadth	-0.10	+0.38
Prosthion-nasion height	+2.59	-0.01
Nasal breadth	+10.56	+2.45

Males: Score > 89.27 = Black; Score < 89.27 = White; Females: Score > 9.22 = Black; Score < 9.22 = White

SOURCE: *Giles and Elliot 1962.*

Over the years, it has become apparent that the need for a database of measurements from forensic cases of known individuals was necessary to aid in the identification of unknown individuals. As such, the Forensic Data Bank was created by Jantz and Moore-Jansen (1988) including information about contemporary forensic cases from around the country. This databank is especially useful since many of the metric methods used to determine sex or ancestry have come from older skeletal collections.

Furthermore, the Forensic Data Bank was used to come up with discriminant functions to better allow investigators to discriminate between "white" and "black" racial categories (see table 2). The creation of the Fordisc computer program in 1993 by Jantz and Ousley opened up a new era for forensic anthropology. This computer program allows researchers to input the standardized measurements and then classifies an unknown individual based on the discriminant function formulae created from the Forensic Data Bank. Thus, ancestry can be determined morphologically by the researcher and then reinforced by the output given from Fordisc. Table 3 shows Fordisc output identifying ancestry.

TABLE 3: *Fordisc output discriminating between white male, black male, and Amerindian male*

FORDISC 2.0 DISCRIMINANT FUNCTION RESULTS USING 21 VARIABLES

GROUP	TOTAL NUMBER	INTO GROUP			PERCENT CORRECT
		WM	BM	AM	
WM	114	112	2	0	98.2 %
BM	88	4	83	1	94.3 %
AM	46	1	2	43	93.5 %
Total: 248			Correct: 238	96.0 %	

MULTIGROUP CLASSIFICATION

GROUP CLASSIFIED INTO	DISTANCE FROM	POSTERIOR PROBABILITIES	TYPICALITY
WM	36.3	.066	.020
BM	32.8	.368	.048
AM ** AM **	32.0	.565	.059

** This case is closest to AMs.

Bare Bones: A Survey of Forensic Anthropology

Group Means		WM 114	BM 88	AM 46
GOL	179	186.8	186.4	179.9
XCB	145	140.8	137.2	143.0
ZYB	141	130.5	131.2	142.0
BBH	137	141.5	132.8	132.9
BNL	103	105.5	101.9	103.2
BPL	94	96.8	102.9	100.7
MAB	67	61.5	66.6	66.5
MAL	50	53.7	57.8	55.0
AUB	124	123.6	120.6	132.2
WFB	99	97.4	97.1	97.1
NLH	57	52.3	51.7	54.0
NLB	25	23.8	26.1	26.0
OBB	41	40.7	39.9	42.9
OBH	37	33.4	34.4	35.4
EKB	102	97.8	100.7	101.9
DKB	26	21.7	24.5	22.5
FRC	105	114.2	111.6	110.7
PAC	116	117.7	118.3	109.5
OCC	97	99.4	95.8	93.8
FOL	35	37.5	36.2	36.5
MDH	35	32.1	32.3	29.4

SOURCE: *Jantz and Ousley 1996.*

In this case, the specimen is categorized as Amerindian.

OTHER ASPECTS OF ANCESTRY DETERMINATION

While the skull is by far the best area to look for assessing population affinity, there are some areas in the postcranial skeleton that are also useful. One of these areas is the amount of curvature in the femur, often referred to as **anterior femoral curvature**. The degree of curvature has been found to differ between ancestral groups. In many African Americans, the femur tends to be straight and lack significant anterior curvature. Native Americans tend towards more torsion in the femoral neck and greater anterior curvature of the shaft, and those of European descent are more often intermediate between the two groups (Stewart 1962).

TABLE 4: *Discriminant function coefficients for distinguishing American whites from American blacks on the basis of cranial measurements*

MEASUREMENT	MALE WEIGHTS	FEMALE WEIGHTS
		...
Maximum width	−0.173441	...
Basion-bregma height	−0.150682	−0.146275
Basion-nasion	−0.276629	−0.552516
Basion-prosthion	0.386880	0.741924
Min. frontal diameter	0.192521	...
Nasal breadth	0.369400	0.746623
Orbital height	0.548942	0.685532
Constant	−11.40607	−38.97745

Score > 0 = Black; Score < 0 White   SOURCE: *Jantz and Moore-Jansen 1988.*

SUMMARY

Determination of ancestry is among the most difficult tasks for the forensic anthropologist. It is not unusual for cases to present mixed indicators of ancestry, making estimation unreliable (and therefore risky).

As practicing anthropologists, we must continue to refine our data so we can continue to include this category of the biological profile in our reporting. Determination of ancestry is necessary in order to estimate stature, since, as we have seen in the previous chapter, different ancestral groups exhibit slightly different limb and body proportions. Therefore, ancestry, or race, remains an important part of our description of the decedent.

Bare Bones: A Survey of Forensic Anthropology

References

Giles, E., and E. Elliot. 1962. Race identification from cranial measurements. *Journal of Forensic Sciences* 7: 147–157.

Gill, G. W. 1998. Craniofacial criteria in the skeletal attribution of race. In *Forensic osteology: Advances in the identification of human remains*, ed. K. J. Reichs, 293–317. 2nd ed. Springfield, IL: Charles C. Thomas.

Hefner, J. T. 2003. Assessing nonmetric cranial traits currently used in the forensic determination of ancestry. Master's thesis, Univ. of Florida.

Jantz, R. L., and P. H. Moore-Jansen. 1988. A database for forensic anthropology: Structure, content, and analysis. Report of Investigations. 47, Department of Anthropology, Univ. of Tennessee at Knoxville.

Jantz, R. L., and S. D. Ousley. [1993-1996] 2005. FORDISC: Personal computer forensic discriminant functions. 3rd vers. Knoxville: University of Tennessee.

Lieberman, L., and F. L. C. Jackson. 1995. Race and three models of human origin. *American Anthropologist* 97:231–242.

Littlefield, A., L. Lieberman, and L. T. Reynolds. 1982. Redefining race: The potential demise of a concept in physical anthropology. *Current Anthropology* 23:641–655.

Montagu, A. 1962. The concept of race. *American Anthropologist* 64:919–928.

Parr, N. M. 2005. Determination of ancestry from discrete traits of the mandible. Master's thesis, Univ. of Indianapolis.

Rathbun, T. A., and J. E. Buikstra. 1984. *Human identification: Case studies in forensic anthropology*. Springfield, IL: Charles C. Thomas.

Rhine, S. 1990. Nonmetric skull racing. In *Skeletal attribution of race: Methods for forensic anthropology*, ed. G.W. Gill and S. Rhine, 7–20. Albuquerque: Maxwell Museum of Anthropology.

Stewart, T. D. 1962. Anterior femoral curvature: Its utility for race identification. *Human Biology* 34:49–62.

Sample Test Questions

1. **If the race concept is not valid as a biological concept, why do forensic anthropologists use it?**

   a. Forensic anthropologists interact with lay persons (the jury) who may be convinced that race is real

   b. Forensic anthropologists work within a social context that has used race as an identifying trait

   c. Forensic anthropologists provide a service. If we can estimate social race, then we should

   d. All of the above are true

2. **Skeletal variation with a "racial category" is often greater than:**

   a. Variation between two different racial categories

   b. One might expect given the genetic complexity of the human species

   c. 10% or more of the overall expected variation

   d. B and C

3. **For most forensic anthropologists, the most diagnostic area for determination of ancestry is:**

   a. The pelvis

   b. The different number of bones between races

   c. The midfacial skeleton

   d. There is no diagnostic area for ancestry determination, because ancestry does not exist

4. **"Race" has been defined variously by:**

   a. Phenotype

   b. Geography

   c. Ethnicity

   d. All of the above

5. **Which of the following cranial traits may be used in the determination of ancestry?**

   a. Shape of the nasal aperture

   b. Shape of the eye orbits

   c. Palatal shape

   d. All of the above

6. **Anthropologists define Hispanic individuals *biologically* by what features?**

   a. The term Hispanic does not define a biological group

   b. The languages they speak

   c. Tooth morphology and cranial shape

   d. A method known as flux capacitating

8

Determination of Age at Death

Age at death is one of the four components of the biological profile. The estimated age offered by the anthropologist will range from tight intervals during the fetal period and early infancy to larger intervals for elderly persons. Investigators use the anthropologist's estimate of age at death to narrow their search among missing persons, to exclude possible missing persons, and as one aspect of presumptive evidence for identity of the decedent.

The human body develops, grows, and matures from conception until adulthood. As time passes, our bodies eventually begin to age and degenerate, as both genetics and the physical aspects of our lives take their toll. The timing of these events—and the myriad ways in which they affect the skeleton—are well documented.

For our purposes, we will divide this continuum of life processes into broad categories of (1) growth and development, (2) maturation, and (3) degeneration. As a fourth category we will examine metamorphic anatomical changes unrelated to growth or degenerative changes. Each of these periods of life offers the anthropologist evidence as to "skeletal age," which is assumed to be closely related to chronological age.

GROWTH AND DEVELOPMENT

From conception until the end of adolescence marks the period of growth and development. During this stage, there exists a good relationship between stature and chronological age, at least prior to the adolescent growth spurt. Therefore, the forensic anthropologist can estimate age at death by calculating the stature of the decedent from conception through around age twelve. Variation in growth rates and the timing of cessation of growth after twelve years of age renders the relationship between age and stature unreliable.

Most of the numerous studies on human growth and development involve longitudinal studies of clinical interest. Analysis of a juvenile's weight and height (or, in the fetal and infancy periods, length) at different stages of development are used to assess growth sufficiency and to detect factors—some of which are pathological—that might contribute to growth acceleration and retardation. Many of these studies are published as standards that establish normal percentiles of growth for a given age. Physicians and other clinicians use standards based on appearance of ossification centers and/or bone morphology to assess skeletal age. Comparison of skeletal age with chronological age provides an indicator of level of maturity, predicted adult height, and other factors related to health-care decisions (Tanner et al. 1983; Greulich and Pyle 1959; Pyle and Hoerr 1969; Roche 1988). Apart from providing standards for growth assessment, anthropologists utilize those same indicators of skeletal age for comparative studies and forensic work (Warren et al. 1997; Jantz and Owsley 1984; Hummert and Van Gerven 1983; Merchant and Ubelaker 1977; Armelagos et al. 1972).

The Fetal Period

Growth in length of long bones is curvilinear during prenatal development, and several standards have been established for both clinical and nonclinical applications (Fazekas and Kósa 1978; Brenner et al. 1976; Birkbeck et al. 1975a, 1975b; Mehta and Singh 1972; Russell et al. 1972; Lubchenco et al. 1966). Most studies of fetal growth are conducted during the first twenty-six weeks of pregnancy because of an increased frequency of spontaneous and elective abortions during the first trimester. (Bagnall et al. 1978, 1982).

Fetal length is usually measured as either total length (called crown-heel length) or as crown-rump length, which is roughly analogous to sitting height. Most studies utilize crown-heel length, because that measurement has been found to contain less interobserver error than crown-rump length (Scammon and Calkins 1925).

The close correlation between total fetal length and long-bone length to period of growth has been used to determine developmental age in both clinical and anthropological contexts. Researchers have used the strong correlation between fetal length and developmental age to establish formulae for estimating developmental age from crown-heel length and long-bone diaphyseal length. **Table 1** shows the remarkable consistency in results correlating lunar months of pregnancy to crown-heel length derived from three early studies. **Haase's rule** is an easily remembered rule of thumb useful for quick calculation of approximate fetal length for each lunar month. Since fetal growth is curvilinear during the first five months, one must simply take the square of each month to arrive at an estimate of crown-heel length. Beginning in the sixth month, one multiplies the lunar month by the number five, since growth during this period is linear—with the fetus growing an absolute length of five centimeters a month. As seen in table 1, this rule of thumb is very close to the results reached by several subsequent researchers (Haase 1877).

TABLE 1: *CHL in cm., with respect to fetal age in lunar months*

LUNAR MONTHS	DIETRICH (1925)	SCAMMON AND CALKINS (1929)	HAASE (1877)
1	. . .	. . .	1.0
2	3.0	. . .	4.0
3	9.8	7.0	9.0
4	18.0	15.5	16.0
5	25.0	22.7	25.0
6	31.5	29.2	30.0
7	37.1	35.0	35.0
8	42.5	40.4	40.0
9	47.0	45.4	45.0
10	50.0	50.2	50.0

SOURCE: *Fazekas and Kósa 1978.*

The gestational age of a fetus may be a primary issue in forensic contexts. The earliest age of biological viability is around twenty-three weeks. However, the *legal* age of viability varies from jurisdiction to jurisdiction and may be as early as 20 weeks. In these jurisdictions, a fetus that is less than twenty weeks in gestational age may be considered biological tissue, but after twenty weeks may be considered a person.

In cases where the fetal remains are decomposed or skeletonized, it becomes necessary to calculate crown-heel length (and thus gestational age) based on the length of the long-bone diaphysis or other skeletal indicators (e.g., Olivier and Pineau 1958; Fazekas and Kósa 1978; Warren 1999). The correlation between fetal long bones and crown-heel length is consistent with that of adult long-bone length and stature: $r2 ? 0.8375$; $p < 0.01$ (Genovés 1967; Trotter and Gleser 1958).

Table 2 shows the linear-regression formulae for predicting the crown-heel length of a fetus using measurements of long-bone diaphyseal length (Olivier and Pineau 1960).

TABLE 2: *Predicting crown-heel length from long bone length*

CHL	=	6.839	(humerus)	+	45.571	±	7.704
CHL	=	8.196	(radius)	+	47.886	±	8.696
CHL	=	7.193	(ulna)	+	51.642	±	8.097
CHL	=	5.188	(femur)	+	90.835	±	7.866
CHL	=	6.308	(tibia)	+	82.858	±	8.351
CHL	=	6.896	(fibula)	+	79.677	±	9.948

For fetal cases involving fleshed, burned, or mummified fetal remains, it is possible to measure the lengths of the bones from radiographs; **table 3** shows the linear-regression formulae for estimating crown-heel length (CHL) from radiographic measurements of long-bone diaphyses (Warren 1999).

Bare Bones: A Survey of Forensic Anthropology

TABLE 3: *Predicting crown-heel length from long-bone radiographic length*

CHL	=	6.839	(humerus)	+	45.571	±	7.704
CHL	=	8.196	(radius)	+	47.886	±	8.696
CHL	=	7.193	(ulna)	+	51.642	±	8.097
CHL	=	5.188	(femur)	+	90.835	±	7.866
CHL	=	6.308	(tibia)	+	82.858	±	8.351
CHL	=	6.896	(fibula)	+	79.677	±	9.948

The Postnatal Period and Childhood

The relationship between long-bone diaphyseal length and stature extends into infancy and childhood. Clinical research has established standardized curves to monitor growth so that infants and children can be treated as early as possible for pathological conditions that accelerate or retard growth velocities. These studies—as with the fetal research discussed above—have enabled anthropologists to estimate age by comparing the length of the long-bone diaphyses of an unknown individual with the lengths measured from normal, healthy children participating in longitudinal growth and development projects. This means we can track the correlation between limb length and stature from birth to arrive at an age estimate for an unknown decedent. **Table 4**, below, gives the average lengths of each of the six long bones between one and twelve years of age (Maresh 1955; after Caffey 1978).

TABLE 4: *Average diaphyseal radiographic long lengths for children aged 1 through 12 years*

AGE, YEARS	1	2	3	4	5	6	7	8	9	10	11	12
Humerus	10.6	13.0	14.8	16.5	17.9	19.3	20.7	22.0	23.2	24.5	25.8	27.0
Radius	7.9	9.7	11.1	12.3	13.1	14.4	15.3	16.3	17.2	18.1	19.0	19.9
Ulna	9.0	10.9	12.4	13.7	14.9	15.9	16.8	17.8	18.7	19.6	20.7	21.7
Femur	13.5	17.1	19.8	22.4	24.8	27.1	29.3	31.5	33.4	35.2	36.8	38.3
Tibia	10.9	14.0	16.3	18.4	20.3	22.1	23.9	25.8	27.5	29.2	30.9	32.6
Fibula	10.5	13.6	16.2	18.2	20.1	21.9	23.7	25.4	27.1	28.7	30.2	31.8

A few anthropological studies have been based on unknown samples (i.e., we do not know their precise ages at death, but we have used a proxy such as "dental age."). These studies include ages two through twelve years, since the investigators had no way of knowing if the children were born at term or were growth impaired. The two years between birth and age two allows for normalization of the growth trajectory by allowing for the common "slow-down" or "catch-up" periods to occur, during which the child's growth curve regresses to the mean (Hoffman 1979).

The Dentition

The development of the dentition is the most consistent indicator of age at death. The development and eruption pattern of the teeth is genetically conservative and less subject to environmental perturbations than the skeleton. Thus, terms such as "six-year molars" and "twelve-year molars" are part of our lexicon and reflect the consistency with which our teeth develop and reach occlusion.

Our first teeth are our deciduous teeth ("baby" teeth or "milk" teeth). These teeth develop in crypts within the jaws in fairly consistent order. Eventually, they erupt and reach occlusion, also within a predictable sequence. Later, the permanent teeth develop and, as they erupt, push the deciduous teeth out.

Researchers and clinicians have produced numerous studies that enable a forensic anthropologist or odontologist to compare the degree of dental development with data from known populations in order to arrive at an estimation of age. Various techniques include observations of the eruption pattern (e.g., Ubelaker 1978), degree and extent of crown-root development (e.g., Demirjian and Levesque 1980; Moorrees et al. 1963a, 1963b), or both (Garn et al. 1958).

Table 5 shows the average age at which the permanent premolars and first and second molars reach occlusion. **Figure 1** presents a postmortem radiograph of a child's mandible showing the developing permanent dentition in the crypts below the deciduous molars.

TABLE 5: *The mean ages of occlusion for the premolars and 1st and 2nd molars*

TOOTH	BOYS' MEAN AGE OF OCCLUSION	GIRLS' MEAN AGE OF OCCLUSION
1st premolar	10.9	10.3
2nd premolar	12.2	11.3
1st molar	6.9	6.9
2nd molar	12.7	11.8

Note that the females are more advanced in development than the males. This is also true of general skeletal development.

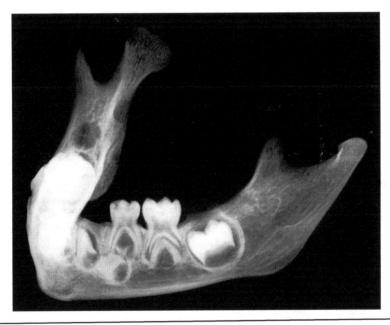

■**FIGURE 1:** *This radiograph shows a mandible with both deciduous molars in occlusion and a first permanent molar developing in its crypt. Calculated age is 4 years ± 12 months of age (Ubelaker 1978), or means 3.7 years for females and 4.1 years for males using the Demirjian and Levesque (1980) method examining crown-root development.*

The Primary Ossification Centers

Bones develop from ossification centers within membranous tissue (dermal, or membranous, bone) or within a cartilaginous template (cartilaginous bone). Once the osseous tissue begins to mineralize, these growth centers develop their final size and shape over a period of years. In the long bones of the extremities, ossification starts near the midshaft of the bone and spreads to either end, with further growth taking place at the physes, or growth plates, situated at the terminal end of the metaphysis. The timing and sequence of ossification among the bones is—like the teeth—very well documented and of great utility for estimating age in both living and deceased individuals. Orthopedists and other clinicians often use radiographs of the wrist and hands to estimate skeletal age for comparison with known chronological age, giving them a starting place from which to diagnosis and treat growth disorders or plan the timing of surgical procedures relative to cessation of growth.

The wrist and hand is an excellent place to look! It has twenty-seven bones to evaluate; the posterior–anterior radiographic view has very little overlap of bones (making each bone easier to evaluate and score); and it is a relatively noninvasive radiograph in terms of limiting a patient's exposure to radiation.

For example, babies born at full term rarely have ossified centers in the wrist. Around the first six months of life, centers for the capitate and hamate appear. Each carpal bone then appears, in a quite predictable sequence, until all are present. Each center then expands to take the shape of its cartilaginous template until the final adult form is reached. Radiographs of the wrist and hands also provide views of the developing secondary centers of the distal radius and ulna, the distal ends of the metacarpals, and the proximal ends of the phalanges. All of these elements must be assessed if one is to provide an accurate age estimate.

Radiographic methods of age estimation can be either quantitative (Tanner et al. 1983; Roche 1988), in which a score is given to access the maturity of a primary center of ossification, or qualitative, in which the investigator compares the patient's (or victim's) hand radiograph with a known standard (Gruelich and Pyle 1959; Pyle and Hoerr 1969). The latter is called an "atlas method," since it uses radiographic atlases of the wrist/hand, shoulder, knee, and ankle/foot for comparison with a specimen.

Bare Bones: A Survey of Forensic Anthropology

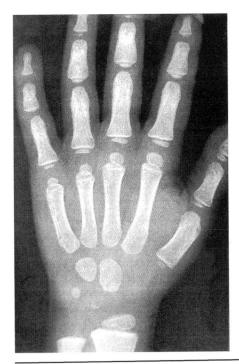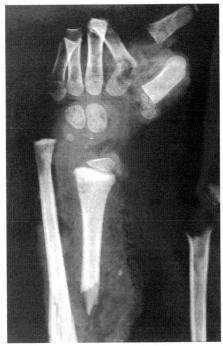

■FIGURE 2A AND 2B: *Hand and wrist standard radiograph of a three-year-old (left) and a postmortem radiograph of the hand of a three-year-old aircraft-crash victim (right). Note the similarity in degree of ossification of the four carpal bones between the radiographic standard and the postmortem radiograph.*

MATURATION

In our categorization scheme, the period of maturation begins when growth ceases at each growth plate or center and the secondary centers begin to fuse to the primary ossification centers.

The Secondary Ossification Centers

The epiphyses lie adjacent to the physis, or growth plate, appearing as secondary ossification centers (**Figure 3**).

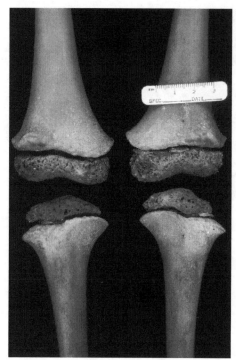

■FIGURE 3: *The epiphyses at the knee joint were still developing and unfused at the time of this juvenile's death.*

As with the primary centers, their appearance and later fusion to the primary center is closely regulated and well documented. Ossification-sequence polymorphisms are relatively rare between populations, so this method is applicable across geographic regions and diverse groups. Only the timing of ossification must be established for each population. For example, in no population is it possible (barring traumatic injury or pathological condition) that the epiphysis of the proximal humerus will fuse to the metaphysis before the epiphysis of the proximal ulna. Fusion of epiphysis to metaphysis begins at the elbow joint, proceeds to the hip joint, then the ankle, then the knee, on to the wrist joint, and finally the shoulder joint. **Figure 4** shows a common diagram and mnemonic device used by anthropologists for remembering the sequence.

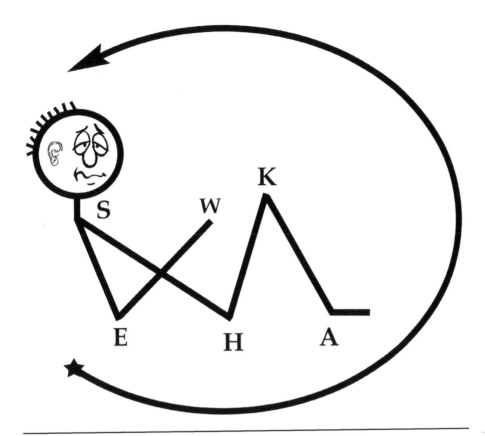

■FIGURE 4: *Even half-witted archaeologists know which shovel!*

Numerous studies have been conducted on the timing of epiphyseal fusion. Several early studies used radiographs of living, known individuals to establish ages for fusion of each epiphysis. A few of these early studies are still useful for practicing forensic anthropologists, particularly when radiographs are being used to assess fusion—as may be the case in mass disasters or genocide. One such study is shown below in **table 6.** Using this chart, a male specimen with a fused femoral head would be at least fourteen years of age, since this is the age at which the youngest individual in Flecker's sample displayed a fused femoral head.

TABLE 6: *Selected epiphyses from Flecker's (1942) original radiographic study (ages are in years and months)*

EPIPHYSES	(A) FEMALE		MALE		(B) FEMALE	MALE	(C) FEMALE		MALE	
Femoral head	13	4	14	0	14	17	18	2	20	2
Distal femur	14	0	16	0	17	19	19	0	19	0
Distal tibia	13	0	14	9	14	17	16	4	18	0
Proximal tibia	14	0	16	0	15	18	18	0	19	0
Humeral head	15	9	16	0	17	18	20	5	19	2
Distal humerus	13	4	14	7	13`	16	16	0	16	0
Medial epicondyle	10	0	12	0	14	16	16	0	17	0
Proximal radius	13	10	14	0	14	16	19	10	20	5
Distal radius	15	11	17	3	18	19	20	5	23	0
Olecranon process	13	10	...		14	16	16	0	17	6
Distal ulna	15	0	17	3	17	19	22	0	23	0
1st metacarpal	13	0	14	0	15	18	15	0	19	7
Acromion process	...		...		17	17	16	1	19	2
Medial clavicle	...		...		18	...	26	11	25	0
Iliac crest	17	1	...		21	21	22	3	22	9
Union of ilium, ischium, and pubis	10	6	13	7	13	15	16	10	17	11

NOTE: *Age is represented as follows: (a) age of individual with earliest fusion, (b) age where 50 percent of the sample shows fusion, (c) age of oldest individual without fusion.*

Other studies were based on examination of dry bone and are most appropriate for cases in which the examiner is grossly examining skeletal material. Since each study may have used a different scoring technique or different normative sample, one must be careful to fully appreciate the possible range of variation across time and populations. It is perhaps the best practice to consider several sources prior to arriving at an age estimate. **Table 7** shows age estimates for fusion of the medial clavicle and iliac crest, derived from a large sample of modern Americans aged eleven to forty years (Owings Webb and Suchey 1985).

TABLE 7: *Fusion chart scoring epiphyses as (1) not ossified, (2) present but separate from the primary center, (3) partially fused, and (4) fused to the primary center*

	CLAVICLES		ILIAC CREST	
	MALE	FEMALE	MALE	FEMALE
None (1)	≤ 25	≤ 23	≤ 16	≤ 11
Separate (2)	16–22	16–21	13–19	14–15
Partial (3)	17–30	16–33	14–23	14–23
Fused (4)	≥ 21	≥ 20	≥ 17	≥ 18

Once the dentition has fully developed and reached occlusion and all of the epiphyses have fused to their primary centers, an individual is fully mature. The last two centers to fuse are the medial epiphyses of the clavicle and the first and second vertebrae of the sacrum. After this stage of life, the body begins a long period of age-related degeneration. Age indicators of degeneration rely on nutritional and health status, occupation, and other unknown factors, so anthropologists must present wider age intervals for decedents who have reached this stage of life.

DEGENERATION

The final phase of our three life phases is the degenerative phase. As the body ages, general bone quality deteriorates. The bones become more porous and less dense. A condition called **osteoporosis** creates light, porous bone susceptible to fracture. Wear and tear on our skeletal framework is manifested as osteoarthritis, an age-related degenerative condition of the joint surfaces. Osteoarthritis shows up as bony extensions and erosion of the articular surfaces of load-bearing bones, such as the bones of the vertebrae and legs (**Figure 5**).

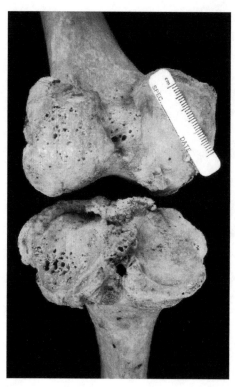

■FIGURE 5: *Degenerative joint disease of the knee.
Note the irregularity of the articular surface.*

A diet high in carbohydrates and processed sugars can cause dental disease such as caries and periodontitis, which can lead to tooth loss. The first and second molars are often lost first, since they have been exposed to pathogens longer than the third molars. The anterior teeth lack the crenulations and crypts of the multicusped molars, so are usually retained longer than the molars and premolars. Tooth loss is seen by most anthropologists as an age indicator—the more missing dentition, the older the individual was at death. Of course, tooth-loss patterns and rates vary with populations, so the investigator must be familiar with the population within his or her area of practice (**Figure 6**).

Bare Bones: A Survey of Forensic Anthropology

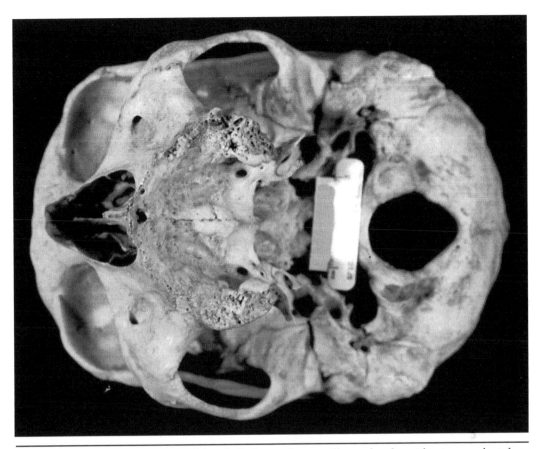

■FIGURE 5: *This basilar view of the skull shows the maxillary alveolar ridge in an edentulous cranium. The ridge has been remodeled by denture pressure.*

Dental wear is another indicator of age. This has been most applicable for pre-historic populations where abrasive, gritty foods were ingested prior to the introduction of processed sugar to the diet. After the introduction of maize agriculture, processing techniques to ground corn introduced another source of grit into the diet. There are several different methods used by bioarchaeologists for scoring tooth wear.

Determination of Age at Death

METAMORPHIC CHANGES

Some anatomical structures seem to change throughout our lives in ways that do not closely correspond with developmental events or age-related pathology. The "face" of the pubic symphysis—the joint in the anterior aspect of the pelvis where the two pubic bones articulate—is an excellent example of a metamorphic process that has proved to be quite accurate in estimating skeletal age. Suchey, Brooks, and various colleagues used a large sample of specimens from known decedents (n > 1100) to develop a phase method of determining age from this anatomical structure (e.g., Katz and Suchey 1986; Brooks and Suchey 1990). This method uses descriptions of the surface of the pubic symphysis, in association with photographs, line drawings, or casts that one may compare with an unknown specimen. Each "phase" corresponds with a mean age, standard deviation, and age interval (see table 8).

TABLE 8: *Means and ranges for Suchey–Brooks's phases*

PHASE	FEMALE (N=273)			MALE (N=739)		
	MEAN	S.D.	95%	MEAN	S.D.	95%
I	19.4	2.5	15–24	18.5	2.1	15–23
II	25.0	4.9	19–40	23.4	3.6	19–34
III	30.7	8.1	21–53	28.7	6.5	21–46
IV	38.2	10.9	26–70	35.2	9.4	23–57
V	48.1	14.6	25–83	45.6	10.4	27–66
VI	60.0	12.4	42–87	61.2	12.2	34–86

Phase I: Symphyseal face has a billowing surface (ridges and furrows), which usually extends to include the pubic tubercle. The horizontal ridges are well marked, and ventral beveling may be commencing. Although ossific nodules may occur on the upper extremity, a key to the recognition of this phase is the lack of delimitation of either extremity (upper and lower).

Phase II: Symphyseal face may still show ridge development. The face has commencing delimitation of lower and/or upper extremities occurring with or without ossific nodules. The ventral rampart may be in beginning phases as an extension of the bony activity at either or both extremities.

Bare Bones: A Survey of Forensic Anthropology

Phase III: Symphyseal face shows lower extremity and ventral rampart in process of completion. There can be a continuation of fusing ossific nodules forming at the upper extremity and along the ventral border. Symphyseal face is smooth or can continue to show distinct ridges. Dorsal plateau is complete. Absence of lipping of symphyseal dorsal margin; no bony ligamentous outgrowths.

Phase IV: Symphyseal face is generally fine-grained, although remnants of the old ridge and furrow system may still remain. Usually the oval outline is complete at this stage, but a hiatus can occur in the upper ventral rim. Pubic tubercle is fully separated from the symphyseal face by definition of upper extremity. Symphyseal face may have distinct rim. Ventrally, bony ligamentous outgrowths may occur on the inferior portion of the pubic bone adjacent to the symphyseal face. If any lipping occurs, it will be slight and located on the dorsal border.

Phase V: Symphyseal face is completely rimmed with some slight depression of the face itself, relative to the rim. Moderate lipping is usually found on the dorsal border with more prominent ligamentous outgrowths on the ventral border. There is little or no rim erosion. Breakdown may occur on the superior ventral border.

Phase VI: Symphyseal face may show ongoing depression as the rim erodes. Ventral ligamentous attachments are marked. In many individuals, the pubic tubercle appears as a separate bony knob. The face may be pitted or porous, giving an appearance of disfigurement with the ongoing process of erratic ossification. Crenulations may occur. The shape of the face is often irregular at this stage.

Figure 7 shows two sets of pubic symphyseal faces. The pair on the left would be scored a phase 5 (mean age 45.6 years for males), while the pair on the right are phase I (mean age 18.5 years for males).

Similar phase methods are used to evaluate the auricular surface of the ilium and the right fourth rib (Buckberry and Chamberlain 2002; Lovejoy et al. 1985; İşcan et al. 1984).

References

Armelagos, G. J., J. H. Meilke, K. H. Owen, D. P. Van Gerven, J. R. Dewey, and P. E. Mahler. 1972. Bone growth and development in prehistoric populations from Sudanese Nubia. *Journal of Human Evolution* 1:89–119.

Bagnall, K. M., P. F. Harris, and P. R. M. Jones. 1978. Studies on the pattern of ossification in the long bones of human fetal limbs with some observations on sex and size differences. *Journal of Bone and Joint Surgery* 60 (B): 284.

——. 1982. A radiographic study of the longitudinal growth of primary ossification centers in long bones of the human fetus. *A natomical Record* 203:293–299.

Birkbeck, J. A., W. A. Billewicz, and A. M. Thomson. 1975a. Human fetal measurements between 50 and 150 days gestation. *Annals of Human Biology* 2:173–178.

——. 1975b. Fetal growth from 50 to 150 days of gestation. *Annals of Human Biology* 2:319–326.

Brenner, W. E., D. A. Edelman, and C. H. Hendricks. 1976. A standard of fetal growth for the United States of America. *American Journal of Obstetrics and Gynecology* 126 (5): 555–564.

Brooks, S., and J. M. Suchey. 1990. Skeletal age determination based on the os pubis: A comparison of the Acsádi-Nemeskéri and Suchey-Brooks methods. *Human Evolution* 5 (3): 227–238.

Buckberry, J. L., and A. T. Chamberlain. 2002. *American Journal of Physical Anthropology* 119 (3): 231–239.

Caffey, J. 1978. *Pediatric X-ray diagnosis*. 7th ed. Vol. 2. Chicago: Year Book Medical.

Demirjian, A., and G. Y. Levesque. 1980. Sexual differences in dental development and prediction of emergence. *Journal of Dental Research* 59:1110–1122.

Dietrich, H. A. 1925. Anatomie und physiologie des fetus und biologie der placenta. *Halban-Seitz I*. Augl. VI, 1:177.

Fazekas, I., and K. Kósa. 1978. *Forensic fetal osteology*. Budapest, Hungary: Akademiai Kiado Publishers.

Garn, S. M., A. B. Lewis, K. Koski, and D. L. Polacheck. 1958. The sex differences in tooth calcification. Journal of Dental Research 37:561–567.

Genovés, S. 1967. Proportionality of the long bones and their relation to stature among Meso-Americans. *American Journal of Physical Anthropology* 26:67–77.

Greulich, W. W., and S. I. Pyle. 1959. *Radiographic atlas of skeletal development of the hand and wrist.* 2nd ed. Stanford, CA: Stanford University Press.

Haase, W. 1877. Maternity annual report for 1875. *Charité Ann* 2:669.

Hoffman, J. M. 1979. Age estimation for diaphyseal lengths: Two months to twelve years. *Journal of Forensic Sciences* 24:461.

Hummert, J. R., and D. P. Van Gerven. 1983. Skeletal growth in the medieval population from Sudanese Nubia. *American Journal of Physical Anthropology* 60:471–478.

İşcan, M. Y., ed. 1989. *Age markers in the human skeleton.* Springfield, IL: Charles C. Thomas.

İşcan, M. Y., S. R. Loth, and R. K. Wrigh. 1984. Metamorphosis at the sternal rib end: A new method to estimate age at death in white males. *American Journal of Physical Anthropology* 65 (2): 147–156.

Jantz, R. L., and D. W. Owsley. 1984. Long bone growth variation among Arikara skeletal populations. *American Journal of Physical Anthropology* 63:13–20.

Johnston, F. E. 1961. Sequence of epiphyseal union in a prehistoric Kentucky population from Indian Knoll. Human Biology 33:66–81.

—. 1962. Growth of the long bones of infants and young children at Indian Knoll. *American Journal of Physical Anthropology* 20:249–254.

—. 1968. Growth of the skeleton in earlier peoples. In *The skeletal biology of earlier human populations*, ed. D.R. Brothwell, 57–66. Oxford: Pergamon Press.

—. 1969. Approaches to the study of developmental variability in human skeletal populations. *American Journal of Physical Anthropology* 31:335–341.

Katz D., and J. M. Suchey. 1986. Age determination of the male os pubis. *American Journal of Physical Anthropology* 69 (4):427–435.

Kósa, F. 1989. Age estimation from the fetal skeleton. In İşcan 1989, 21–54.

Lovejoy, C. O., R. S. Meindl, T. R. Pryzbeck, and R. P. Mensforth. 1985.

Chronological metamorphosis of the auricular surface of the ilium: A new method for the determination of adult skeletal age at death. *American Journal of Physical Anthropology* 68 (1): 15–28.

Lubchenco, L. O., C. Hansman, and E. Boyd. 1966. Intrauterine growth in length and head circumference as estimated from live births at gestational ages from 26 to 42 weeks. *Pediatrics* 37 (3): 403–408.

Maresh, M. M. 1955 Linear growth of the long bones of extremities from infancy through adolescence. *American Journal of Diseases of Children* 89:725.

Mehta, L., and H. M. Singh. 1972. Determination of crown-rump length from fetal long bones: Humerus and femur. *American Journal of Physical Anthropology* 36:165–168.

Mensforth, R. P. 1985. Relative tibia long bone growth in the Libben and Bt-5 prehistoric skeletal populations. *American Journal of Physical Anthropology* 68:247–262.

Merchant, V. L., and D. H. Ubelaker. 1977. Skeletal growth of the protohistoric Arikara. *American Journal of Physical Anthropology* 46 (1): 61–72.

Moorrees, C. F. A., E. A. Fanning, and E. E. Hunt Jr. 1963a. Formation and resorption of three deciduous teeth in children. *American Journal of Physical Anthropology* 21(2):205–213.

———. 1963b. Age variation of formation stages for ten permanent teeth. *Journal of Dental Research* 42:1490–1502.

Olivier, G., and H. Pineau. 1958. Determination de l'age du foetus et de l'embryon. *Archives of Anatomy (La Semaine des Hopitaux)* 6: 21–28.

———. 1960. Nouvelle determination de la taille foetale d'apres longueurs diaphysaires des os longs. *Annals of Legal Medicine* 40:141–144.

Owings Webb, P. A., and J. M. Suchey. 1985. Epiphyseal union of the anterior iliac crest and medial clavicle in a modern multiracial sample of American males and females. *American Journal of Physical Anthropology* 68 (4): 457–466.

Pyle S. I., and N. L. Hoerr. 1969. *A radiologic standard of reference for the growing knee.* Springfield, IL: Charles C. Thomas.

Roche, A. F. 1988. *Assessing the skeletal maturity of the hand-wrist: Fels method.* Springfield, IL: Charles C. Thomas.

Russell, J. G., A. E. Mattison, and W. T. Easson. 1972. Skeletal dimensions as an indication of foetal maturity. *British Journal of Radiology* 45:667.

Scammon, R. E., and L. A. Calkins. 1925. Crown-heel and crown-rump length in the fetal period and at birth. *Anatomical Record* 29:372–373.

Tanner, J. M., R. H. Whitehouse, N. Cameron, W. A. Marshall, M. J. R. Healy, and H. Goldstein. 1983. *Assessment of skeletal maturity and prediction of adult height (TW2 method).* 2nd ed. San Diego: Academic Press.

Ubelaker, D. H. 1989. The estimation of age at death from immature human bone. In Íşcan 1989, 55–70.

Bare Bones: A Survey of Forensic Anthropology

Warren, M. W. 1999. Radiographic determination of developmental age in fetuses and stillborns. *Journal of Forensic Sciences* 44 (4): 708–712.

Warren, M. W., K. R. Smith, P. R. Stubblefield, S. S. Martin, and H. A. Walsh-Haney. 2000. Use of radiographic atlases in a mass fatality. *Journal of Forensic Sciences* 45 (2): 467–470.

Weaver, D. S. 1986. Forensic aspects of fetal and neonatal skeletons. In *Forensic osteology: Advances in the identification of human remains*, ed. K. J. Reichs, 90–100. Springfield, IL: Charles C. Thomas.

Sample Test Questions

1. You are presented with an intact skeleton, including the skull, of what appears to be a child. What methods might be used to determine age at death?

   a. Check epiphyseal union

   b. Check tooth eruption

   c. Measure long bones

   d. All of the above

2. Using Haase's Rule, what is the crown-heel length of a fetus at 4 lunar months gestation?

   a. 8 centimeters

   b. 8 inches

   c. 16 centimeters

   d. 21 inches

3. There is a good relationship between stature and age up until about the age of:

   a. 6 months

   b. 8 years

   c. 12 years

   d. 18 years

4. The epiphyses of which joints fuse first:

   a. shoulder

   b. elbow

   c. wrist

   d. ankle

Determination of Stature

Stature, or standing height, is the fourth group characteristic that can be derived from measurement of the skeleton. Along with determination of sex, ancestry, and age at death, estimation of stature gives investigators another characteristic to consider when searching among missing-persons records, as well as another line of presumptive evidence in considering whether a set of skeletal remains is consistent with known information for a putative decedent. In addition, estimation of stature in fetuses, infants, and children is closely correlated with age, so can be used to estimate age at death in juvenile cases.

It has long been recognized that basic body proportions are quite consistent from person to person, so that the linear measurement *foot* relates to the "normal" length of a man's foot, and a *yard* roughly describes the length from the end of one's outstretched hand to his nose. Leonardo da Vinci famously captured the symmetry and regularity of body and limb proportions in his drawing *The Vitruvian Man* (**figure 1**). Note how the outstretched hands approximate the man's height. It is this predictable relationship between limb length and stature that enables one to estimate overall height from the length of the long bones.

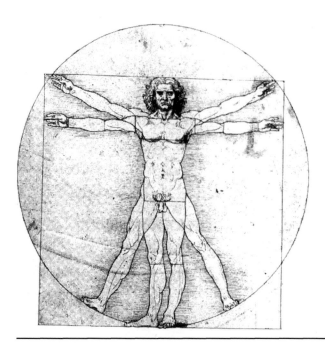

■FIGURE 1: *Leonardo DaVinci's The Vitruvian Man.*

Human body and limb proportions are under tight genetic control and function to balance the need for efficient bipedal locomotion with a body type that is well-adapted to the environment, particularly temperature and terrain. With few exceptions, human populations comply with the ecological rules of Bergmann and Allen; that is, within broadly dispersed geographical populations, both body mass and relative limb length exhibit a clinal pattern with climate and geography, the ultimate effect of which is a significant positive relationship of the body surface area: body mass ratio with temperature (Bergmann 1847; Allen 1877). For humans, those populations from nearer the equator exhibit relatively longer appendages than populations nearer the poles. Moreover, the difference is primarily found in the distal extremities. Therefore, prior to determination of stature, the investigator must make an assessment of ancestry, since populations from different geographic areas of the world exhibit slightly different, but statistically significant, body and limb proportions. In cases where ancestry cannot be confidently estimated, a wider range of stature, inclusive of all ancestral categories, may be reported.

Stature increases from conception until adulthood and then decreases slightly with advanced age. Humans vary greatly in height. Each individual is born

Bare Bones: A Survey of Forensic Anthropology

with a genetic potential for a given stature, and that potential requires optimal nutrition and health to achieve. Studies of identical twins have suggested that 90 percent of stature is genetic and 10 percent is environmental—the difference between the maximum genetic potential and the actual, realized height.

In humans, males are, as a group, taller than females—another aspect of sexual dimorphism discussed in earlier chapters. Males and females also have slightly different limb proportions. As with ancestry, the anthropologist must know the biological sex of the decedent before applying the appropriate stature formulae.

METHODS OF DETERMINING STATURE

Anthropologists employ two primary methods of determining **forensic stature**. The first is referred to as Fully's method, named after the researcher who first described the technique (Lundy 1988; Fully 1956; Fully and Pineau 1960). Fully's method uses measurements of cranial height, the height of the vertebral bodies of the second cervical vertebra through the first sacral vertebra, the physiological lengths of the femur and tibia, and the articulated height of the talus and calcaneus. Then, the following formula is used:

Stature (cm.) = sum of the skeletal elements + 10.8 ± 2.015 cm.

The constant allows for the intervertebral disc spaces. This method requires that all elements be present, so it is not useful when working with incompletely recovered skeletal remains.

The second method involves first measuring one or more long bones, then using a simple linear-regression formula to calculate statute. The maximum length of any of the long bones—the humerus, radius, ulna, femur, tibia, or fibula—may be used alone or in combination (e.g., femur + tibia) to estimate total stature. Long-bone measurements are recorded using an **osteometric board (figure 2)**. Landmark studies in stature determination include those of Trotter and Gleser, beginning in 1952 (Trotter and Gleser 1951, 1952, 1958, 1977). The researchers used a large sample (n = ≈ 5,000) comprised of males and females of European and African ancestry—later adding data for Mexicans and "Mongoloid" males. Their regression formulae are in **table 1**. These equations were derived from data collected from the identified skeletons of 790 males from World War II, 3,382 males from the Korean War, and 615 males and 240 females from the Terry Collection. By using the regression formulae and running every possible iteration, it is easy to develop a stature table for quick reference. Tables 2–5 on the following pages are helpful for providing a quick assessment of stature in the field.

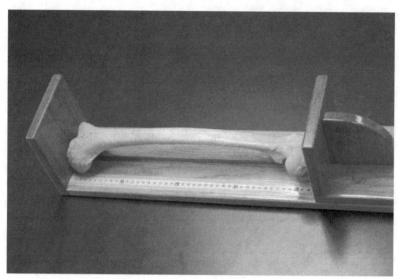

■FIGURE 2: *A simple wooden osteometric board provides accurate measurements of maximum long-bone length.*

Many anthropologists use stature formulae derived from the Forensic Data Bank at the University of Tennessee for forensic cases. These data are taken from measurements of identified forensic cases from around the country, as well as skeletons donated to the William Bass Donated Collection in Knoxville. The formulae are found in the Fordisc computer program and are convenient and accurate (Jantz and Ousley [1993-1996] 2005). Since these skeletons may better represent our contemporary population than the veterans of WWII and the Korean War and the older, cadaver-room populations of the Terry and Hamann-Todd collections, these formulae may be more appropriate as a reference sample for current forensic work.

Problems with Stature Estimates

How tall are you? You probably have an idea based on a measurement taken at the physician's office, or perhaps your military record. Maybe you have had a friend measure you by using a book and marking your height on a door frame. You will find that your recorded stature will vary by up to several inches depending on several factors. We are taller in the morning, when our muscles are relaxed, than in the evening, when we are tired and gravity has done its work. So the time of day will have some impact on measured height. If stature is measured by a trained health-care worker or technician, then results will be more accurate and consistent. The most accurate measurements tend to be those taken by military physicians, who use a prescribed method that dictates correct posture, lack of footwear, and measuring equipment.

It is also known that self-reported stature is significantly greater than measured height, particularly for males (Willey and Falsetti 1991; Giles and Hutchinson 1991). Similarly, **cadaver stature** may vary from **living stature** by as much as two inches.

TABLE 1: *Mildred Trotter's original stature formulae from 1970*

WHITE MALES						BLACK MALES					
3.08	Hum	+	70.45	±	4.05	3.26	Hum	+	62.10	±	4.43
3.78	Rad	+	79.01	±	4.32	3.42	Rad	+	81.56	±	4.30
3.70	Ulna	+	74.05	±	4.32	3.26	Ulna	+	79.29	±	4.42
2.38	Fem	+	61.41	±	3.27	2.11	Fem	+	70.35	±	3.94
2.52	Tib	+	78.62	±	3.37	2.19	Tib	+	86.02	±	3.78
2.68	Fib	+	71.78	±	3.29	2.19	Fib	+	85.65	±	4.08
WHITE FEMALES						BLACK FEMALES					
3.36	Hum	+	57.95	±	4.45	3.08	Hum	+	64.67	±	4.25
4.74	Rad	+	54.93	±	4.24	3.67	Rad	+	71.79	±	4.59
4.27	Ulna	+	57.76	±	4.30	3.31	Ulna	+	75.38	±	4.83
2.47	Fem	+	54.10	±	3.72	2.28	Fem	+	59.76	±	3.41
2.90	Tib	+	61.53	±	3.66	2.45	Tib	+	72.65	±	3.70
2.93	Fib	+	59.61	±	3.57	2.49	Fib	+	70.90	±	3.80
MONGOLOID MALES						MEXICAN MALES					
2.68	Hum	+	83.19	±	4.25	2.92	Hum	+	73.94	±	4.24
3.54	Rad	+	82.00	±	4.60	3.55	Rad	+	80.71	±	4.04
3.48	Ulna	+	77.45	±	4.66	3.56	Ulna	+	74.56	±	4.05
2.15	Fem	+	72.57	±	3.80	2.44	Fem	+	58.67	±	2.99
2.39	Tib	+	81.45	±	3.27	2.36	Tib	+	80.62	±	3.73
2.40	Fib	+	80.56	±	3.24	2.50	Fib	+	75.44	±	3.52

SOURCE: *Bennett 1993.*

TABLE 2: *Estimated stature based on maximum long bone length for American black females, from Trotter (1970)*

HUM MM	RAD MM	ULNA MM	STATURE IN*	STATURE MM	FEM MM	TIB MM	FIB MM
245	186	195	55^1	140	352	275	278
248	189	198	55^4	141	356	279	282
251	191	201	55^7	142	361	283	286
254	194	204	56^2	143	365	287	290
258	197	207	56^6	144	369	291	294
261	199	210	57^1	145	374	295	298
264	202	213	57^4	146	378	299	302
267	205	216	57^7	147	383	303	306
271	208	219	58^2	148	387	308	310
274	210	222	58^5	149	391	312	314
277	213	225	59^0	150	396	316	318
280	216	228	59^4	151	400	320	322
284	218	231	59^7	152	405	324	326
287	221	235	60^2	153	409	328	330
290	224	238	60^5	154	413	332	334
293	227	241	61^0	155	418	336	338
297	229	244	61^3	156	422	340	342
300	232	247	61^6	157	426	344	346
303	235	250	62^2	158	431	348	350
306	238	253	62^5	159	435	352	354
310	240	256	63^0	160	440	357	358
313	243	259	63^3	161	444	361	362
316	246	262	63^6	162	448	365	366
319	249	265	64^1	163	453	369	370
322	251	268	64^5	164	457	373	374
326	254	271	65^0	165	462	377	378
329	257	274	65^3	166	466	381	382
332	259	277	65^6	167	470	385	386
335	262	280	66^1	168	475	389	390

Bare Bones: A Survey of Forensic Anthropology

HUM MM	RAD MM	ULNA MM	STATURE		FEM MM	TIB MM	FIB MM
			IN*	MM			
339	265	283	66^4	169	479	393	394
342	268	286	66^7	170	484	397	398
345	270	289	67^3	171	488	401	402
348	273	292	67^6	172	492	406	406
352	276	295	68^1	173	497	410	410
355	279	298	68^4	174	501	414	414
358	281	301	68^7	175	505	418	418
361	284	304	69^2	176	510	422	422
365	287	307	69^5	177	514	426	426
368	289	310	70^1	178	519	430	430
371	292	313	70^4	179	523	434	434

SOURCE: *Bennett 1993.*

TABLE 3: *Estimated stature based on maximum long bone length for American black males, from Trotter (1970)*

HUM MM	RAD MM	ULNA MM	STATURE		FEM MM	TIB MM	FIB MM
			IN*	MM			
276	206	223	59^7	152	387	301	303
279	209	226	60^2	153	391	306	308
282	212	229	60^5	154	396	310	312
285	215	232	61^0	155	401	315	317
288	218	235	61^3	156	406	320	321
291	221	238	61^6	157	410	324	326
294	224	242	62^2	158	415	329	330
297	226	245	62^5	159	420	333	335
300	229	248	63^0	160	425	338	339
303	232	251	63^3	161	430	342	344
306	235	254	63^6	162	434	347	349
310	238	257	64^1	163	439	352	353
313	241	260	64^5	164	444	356	358

HUM MM	RAD MM	ULNA MM	STATURE IN*	STATURE MM	FEM MM	TIB MM	FIB MM
316	244	263	65^0	165	449	361	362
319	247	266	65^3	166	453	365	367
322	250	269	65^6	167	458	370	371
325	253	272	66^1	168	463	374	376
328	256	275	66^4	169	468	379	381
331	259	278	66^7	170	472	383	385
334	262	281	67^3	171	477	388	390
337	264	284	67^6	172	482	393	394
340	267	287	68^1	173	487	397	399
343	270	291	68^4	174	491	402	403
346	273	294	68^7	175	496	406	408
349	276	297	69^2	176	501	411	413
352	279	300	69^5	177	506	415	417
356	282	303	70^0	178	510	420	422
359	285	306	70^1	179	515	425	426
362	288	309	70^7	180	520	429	431
365	291	312	71^2	181	525	434	435
368	294	315	71^5	182	529	438	440
371	297	318	72^0	183	534	443	445
374	300	321	72^4	184	539	447	449
377	302	324	72^7	185	544	452	454
380	305	327	73^2	186	548	456	458
383	308	330	73^5	187	553	461	463
386	311	333	74^0	188	558	466	467
389	314	336	74^3	189	563	470	472
392	317	340	74^6	190	567	475	476
395	320	343	75^2	191	572	479	481

SOURCE: *Bennett 1993.*

TABLE 4: *Estimated stature based on maximum long bone length for American white females, from Trotter (1970)*

HUM MM	RAD MM	ULNA MM	STATURE IN*	STATURE MM	FEM MM	TIB MM	FIB MM
244	179	193	55^1	140	348	271	274
247	182	195	55^4	141	352	274	278
250	184	197	55^7	142	356	277	281
253	186	200	56^2	143	360	281	285
256	188	202	56^6	144	364	284	288
259	190	204	57^1	145	368	288	291
262	192	207	57^4	146	372	291	295
265	194	209	57^7	147	376	295	298
268	196	211	58^2	148	380	298	302
271	198	214	58^5	149	384	302	305
274	201	216	59^0	150	388	305	309
277	203	218	59^4	151	392	309	312
280	205	221	59^7	152	396	312	315
283	207	223	60^2	153	400	315	319
286	209	225	60^5	154	404	319	322
289	211	228	61^0	155	409	322	326
292	213	230	61^3	156	413	326	329
295	215	232	61^6	157	417	329	332
298	217	235	62^2	158	421	333	336
301	220	237	62^5	159	425	336	340
304	222	239	63^0	160	429	340	343
307	224	242	63^3	161	433	343	346
310	226	244	63^6	162	437	346	349
313	228	246	64^1	163	441	350	353
316	230	249	64^5	164	445	353	356
319	232	251	65^0	165	449	357	360
322	234	253	65^3	166	453	360	363
324	236	256	65^6	167	457	364	366
327	239	258	66^1	168	461	367	370
330	241	261	66^4	169	465	371	373

HUM MM	RAD MM	ULNA MM	STATURE IN*	STATURE MM	FEM MM	TIB MM	FIB MM
333	243	263	66^7	170	469	374	377
336	245	265	67^3	171	473	377	380
339	247	268	67^6	172	477	381	384
342	249	270	68^1	173	481	384	387
345	251	272	68^4	174	485	388	390
348	253	275	68^7	175	489	391	394
351	255	277	69^2	176	494	395	397
354	258	279	69^5	177	498	398	401

SOURCE: *Bennett 1993.*

TABLE 5: *Estimated stature based on maximum long bone length for American white males, from Trotter (1970)*

HUM MM	RAD MM	ULNA MM	STATURE IN*	STATURE MM	FEM MM	TIB MM	FIB MM
265	193	211	59^7	152	381	291	299
268	196	213	60^2	153	385	295	303
271	198	216	60^5	154	389	299	307
275	201	219	61^0	155	393	303	311
278	204	222	61^3	156	398	307	314
281	206	224	61^6	157	402	311	318
284	209	227	62^2	158	406	315	322
288	121	230	62^5	159	410	319	326
291	214	232	63^0	160	414	323	329
294	217	235	63^3	161	419	327	333
297	220	238	63^6	162	423	331	337
301	222	240	64^1	163	427	335	340
304	225	243	64^5	164	431	339	344
307	228	246	65^0	165	435	343	348
310	230	249	65^3	166	440	347	352
314	233	251	65^6	167	444	351	355

Bare Bones: A Survey of Forensic Anthropology

HUM MM	RAD MM	ULNA MM	STATURE IN*	STATURE MM	FEM MM	TIB MM	FIB MM
317	235	254	66^1	168	448	355	359
320	238	257	66^4	169	452	359	363
323	241	259	66^7	170	456	363	367
327	243	262	67^3	171	461	367	370
330	246	265	37^6	172	465	371	374
333	249	267	68^1	173	469	375	378
336	251	270	68^4	174	473	379	381
339	254	273	68^7	175	477	383	385
343	257	276	69^2	176	482	386	389
346	259	278	69^5	177	486	390	393
349	262	281	70^1	178	490	394	396
352	265	284	70^4	179	494	398	400
356	267	286	70^7	180	498	402	404
359	270	289	71^2	181	503	406	408
362	272	292	71^5	182	507	410	411
365	275	294	72^0	183	511	414	415
369	278	297	72^4	184	515	418	419
372	280	300	72^7	185	519	422	422
375	283	303	73^2	186	524	426	426
378	286	305	73^5	187	528	430	430
382	288	308	74^0	188	532	434	434
385	291	311	74^3	189	536	438	437
388	294	313	74^6	190	540	442	441
391	296	316	75^2	191	545	446	445

SOURCE: *Bennett 1993.*

How does advancing age affect stature? Over time, the intervertebral disc spaces of the vertebral column are reduced, and the vertebral centra lose height due to compression—a burden for us bipedal humans. Age-related pathology, such as osteoporosis, and/or traumatic injury, can worsen the compression of the vertebral column. Therefore, anthropologists must make corrections by subtracting from the calculated stature derived from long-bone lengths (Giles 1991). Table 6 shows how much stature should be subtracted due to the aging process.

TABLE 6: *Adjustment for decreasing stature with age (in inches)*

AGE	MALE	FEMALE	AGE	MALE	FEMALE
46	0.1	0.0	66	0.7	0.6
47	0.1	0.0	67	0.7	0.6
48	0.1	0.0	68	0.8	0.7
49	0.1	0.0	69	0.8	0.7
50	0.2	0.0	70	0.9	0.8
51	0.2	0.0	71	0.9	0.9
52	0.2	0.0	72	1.0	0.9
53	0.2	0.0	73	1.0	1.0
54	0.3	0.1	74	1.1	1.1
55	0.3	0.1	75	1.1	1.1
56	0.3	0.1	76	1.2	1.2
57	0.4	0.2	77	1.2	1.3
58	0.4	0.2	78	1.3	1.4
59	0.4	0.2	79	1.3	1.4
60	0.5	0.3	80	1.4	1.5
61	0.5	0.3	81	1.5	1.6
62	0.5	0.4	82	1.5	1.7
63	0.6	0.4	83	1.6	1.8
64	0.6	0.5	84	1.6	1.8
65	0.6	0.5	85	1.7	1.9

SOURCE: *Giles 1991.*

Obviously, the exact age of an unidentified decedent will not be known. However, the estimated age range will give the anthropologist some idea as to whether it is appropriate to consider the extent to which age has impacted stature in a given case. Examination of age-related changes in the vertebral column—which are also considered in estimating age at death—will also serve as an indicator of the extent to which stature may have decreased with advancing age.

Bare Bones: A Survey of Forensic Anthropology

STATURE FROM FRAGMENTARY LONG BONES

It is not uncommon to encounter skeletal remains in which the long bones have been fragmented or scavenged. In these cases, it is necessary to take measurements of the fragmentary bones and calculate a long-bone length. Once overall length is calculated, a standard stature formula may be employed to estimate total stature.

Fortunately, the anatomical structures and landmarks on bones are consistent between individuals. The anthropologist can take measurements between any available landmarks on the fragmented bone and use established linear-regression formulae to estimate total long-bone length (Simmons et al. 1990; Steele 1970; Steele and McKern 1967). For example, a measurement taken from the most distal point on the humeral head to the proximal margin of the olecranon fossa (segment 2 between landmark 2 and 3) would use the following formula: 3.42 (segment 2) + 80.94 cm. ± 5.31 cm. (See **Figure 3**).

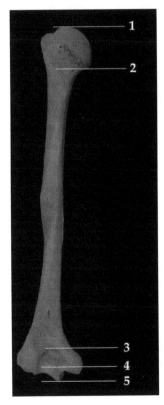

■FIGURE 3: *An image showing landmarks used in estimation of the maximum length of the humerus; Steele 1970).*

Stature Estimation in Subadults

As seen in the preceding chapter, there is a close correlation between a child's height and age. It is usually possible, particularly for those with children at home, to see a child in the supermarket and guess that child's age within a year or two. This correlation is strongest during the fetal period and then grows less strong as the years pass. Once children reach puberty, the relationship between their height and age becomes unreliable, until of course, they reach adulthood, when there is no longer a relationship between height and age. Therefore, measurement of maximum long-bone diaphyseal length and subsequent calculation of stature provides an accurate means of age estimation up until about twelve years of age. Stature estimation in children (or crown-heel length in fetuses) is accomplished in the same way as in adults, except that the long-bone measurements include only diaphyseal length of the bone, because the epiphyses have yet to fuse to the shaft (see Chapter 8).

References

Allen, J. A. 1877. The influence of physical conditions on the genesis of species. *Radical Rev.* 1:108–140.

Bennett, K. A. 1993. *A field guide for human skeletal identification.* Springfield, IL: Charles C. Thomas.

Bergmann, C. 1847. Uber die verhaltniesse der warmeokonomie der thiere zu ihrer grosse. *Gottingen Studien* 1:595–708.

Fully, G. 1956. Une nouvelle méthode de détermination de la taille. *Annals of Médines Légale* 36:266–273.

Fully, G., and H. Pineau. 1960. Détermination de la stature an moyen du squelette. *Annals of Médines Légale* 40:3–11.

Giles, E. 1991. Corrections for age in estimating older adult's stature from long bones. *Journal of Forensic Sciences* 36 (3): 898–901.

Giles, E., and D. L. Hutchinson. 1991. Stature- and age-related bias in self-reported stature. *Journal of Forensic Sciences* 36 (3): 765–780.

Jantz, R. L., and S. D. Ousley. [1993-1996] 2005. FORDISC: Personal computer forensic discriminant functions. 3rd vers. Knoxville: University of Tennessee.

Lundy, J. K. 1988. A report on the use of Fully's anatomical method to estimate stature in military skeletal remains. *Journal of Forensic Sciences* 33 (2): 534–539.

Simmons, T., R. L. Jantz, and W. M. Bass. 1990. Stature estimation from fragmentary femora: A revision of the Steele method. *Journal of Forensic Sciences* 35 (3): 628–636.

Steele, D. G., and T. W. McKern. 1967. A method for assessment of maximum long bone length and living stature from fragmentary long bones. *American Journal of Physical Anthropology* 31 (2): 215–227.

Trotter, M., and G. C. Gleser. 1951. The effect of aging on stature. *Journal of Physical Anthropology* 9 (3): 311–324.

—. 1952. Estimation of stature from long bones of American Whites and Negroes. *American Journal of Physical Anthropology* 10 (4): 463–514.

—. 1958. A re-evaluation of estimation of stature based on measurements of stature taken during life and of long bones after death. *American Journal of Physical Anthropology* 16 (1): 79–123.

—. 1977. Corrigenda to "estimation of stature from long limb bones of American Whites and Negroes." *American Journal Physical Anthropology* 47 (2): 355–356.

Willey, P., and T. Falsetti. 1991. Inaccuracy of height information on driver's licenses. *Journal of Forensic Sciences* 36 (3): 813–819.

Sample Test Questions

1. **Studies have shown that people tend to report their stature:**

   a. Taller than their actual stature

   b. Shorter than their actual stature

   c. Accurately, self-reported stature is usually identical to measured stature

   d. In centimeters as opposed to inches

2. **The best single bone to use in estimating stature is:**

   a. Femur

   b. Humerus

   c. Tibia

   d. Calcaneus

3. **What type of statistical technique is used to determine stature from long bones?**

   a. Simple linear regression

   b. Multiple regression

   c. Analysis of variance (ANOVA)

   d. Discriminant function

4. **Long bone length is measured with:**

   a. An osteometric board

   b. Spreading calipers

   c. Sliding calipers

   d. Dial calipers

10

Personal Identification

Humans develop their personal identities based on a collage of biological and cultural factors. Biological markers include physical traits that place an individual into one or more groups, such as sex, height, skin color, and age. Cultural factors that contribute to a person's identity include an actual or perceived affiliation with a particular set of groups based on gender, self-identified race, ethnicity, language, religion, and socioeconomic status. These **group characteristics**, discussed in following chapters, provide investigators with a biological profile. This **biological profile** is a starting point for law enforcement and other investigators in narrowing down the number of possible decedents from missing-persons lists.

Some traits are unique to each individual and differentiate them from all others. These unique traits, along with both real and perceived group affiliations, constitute one's persona – the way in which an individual is viewed by themselves and others. Personal identity is an integral part of human life and impacts almost every aspect of our relationship with other humans and institutions in our society. Proof of the importance of personal identity is found in the human capacity to store into memory, and immediately recognize, the faces of thousands of people. This chapter will discuss how these unique traits are used to establish personal identity from skeletal remains.

Because personal identity is vital to the human condition, identification of the dead is seen as an important and worthwhile endeavor in almost every culture. Establishing the identity of a decedent permits family and friends to properly grieve for the loss of their loved one; and to exercise customary funerary rites. In societies with complex legal systems, personal identification of the dead is also necessary in order to secure official documents related to legal and financial considerations - including, but not limited to - probate and inheritance, insurance, and contractual obligations. In the legal forum, personal identification of homicide victims may further the cause of prosecution of the guilty.

The identified body is the *corpus delicti* and represents the most basic biological evidence that a crime has been committed. While homicides have been prosecuted without an identified body of the victim, it is of utmost importance for homicide investigators to determine the identity of the decedent in order to establish the relationship between the victim and the offender. The courts, as well as other social organizations with a vested financial interest, are understandably rigid in insisting on proof of death.

Personal identification in living persons is primarily established by general appearance. Therefore, identification of decedents in which the overall appearance of the body has not been altered may be done in the same way in which we identify those people we know: facial form, general build and stature, eye color, hair color and style, and other equally subtle identifying factors. In cases in which the body is either burned, decomposed, or too highly fragmented or disfigured to permit identification by normal means, identification must be accomplished by scientific determination of the decedent's biological profile; and the discovery of unique life history markers and idiosyncratic anatomical variation. Personal identification from the skeleton begins with populational data to place the individual into one of several groups, as seen in previous chapters. Once we know this demographic biological profile, investigators can see if individuals known to be missing are among the possibilities. Investigators then collect personal data for comparison with information gleaned from skeletal analysis, in order to discover if the decedent is indeed the missing person, to the exclusion of all other people.

Positive and presumptive identification

Identification of the dead is based on positive and/or presumptive evidence. **Positive evidence** constitutes the most rigorous scientific standard, approaching 100% certainty. This degree of certainty is the goal in every forensic case and is almost always applied when there is a single decedent of unknown identity at the time of the discovery of the body. Standard methods

Bare Bones: A Survey of Forensic Anthropology

of positive identification, such as fingerprints and nuclear DNA, generally withstand the scrutiny of both the legal and scientific communities and are accepted as absolute biological proof of identity.

Presumptive evidence uses logic to associate remains with a specific individual. Investigators assign probative value to presumptive evidence based on one or more assumptions which are considered within a known **context**. For example, determination of sex and age at death may have an extremely high probative value in identification of the victims of a small-aircraft accident, if the three passengers were an adult male, adult female and child. The assumption here is that the passenger manifest is accurate and we are provided with (and trust) the demographic data for all three passengers on board. However, those same group characteristics would have extremely low probative value in human-rights work involving the examination of skeletal remains of multiple decedents of a single sex, from the same age category and within the same ancestral or ethnic group.

In some contexts, practical realities may lead pathologists and anthropologists to identify a decedent based on the less-rigorous standard of presumptive evidence. In some cases, the probative value of several avenues of presumptive evidence may be so high as to constitute an identification, while if considered alone, each piece of the puzzle may be insufficient to identify the decedent.

Presumptive evidence may include personal effects found at the scene, such as a wallet with documents, or distinctive jewelry. While generally meaningless to most forensic scientists who demand biological proof of identity, these pieces of presumptive evidence have been considered adequate to establish identity in special circumstances. For example, in Srebrennica, a rural area of Bosnia-Herzegovina, the victims of ethnic cleansing had few medical or dental records available for antemortem-postmortem comparison. Therefore, forensic investigators working to identify the massive number of victims of this conflict in some cases deemed recognizable clothing or identifying documents found on the body as proof of identity.

Additional lines of presumptive evidence include comparable life-history events. The value of these lines of evidence follows the same principles as the examples for group characteristics given above. For example, the known information for a missing person may include a history of a broken left forearm. If the unknown skeletal remains exhibit a healed fracture of the left radius, then that would constitute one line of presumptive evidence and would be assigned appropriate probative value. If the decedent is known to be one of only two individuals, such as the pilot or radio control officer of a military aircraft – and only one of those individuals had a history of a fractured radius—then that same piece of biological evidence is assigned much greater probative value.

There is ongoing debate among scientists as to what constitutes a "good" identification. Some practitioners use the terms "positive ID" and "presumptive ID" depending on the types of evidence used. This may lead some to conclude that a "presumptive ID" is less convincing than a "positive ID." We encourage our colleagues to simply use the term "identification", then, if needed, they may specify whether positive or presumptive lines of evidence were used to reach their conclusion (Steadman *et al.*, 2006; Anderson, 2007; Steadman *et al.*, 2007).

Biological factors in personal identification from the skeleton

Biological factors for identification from the skeleton were outlined by C.P. Warren (1978) and include *idiosyncratic morphology* and both physical and cultural *life-history variables*. Every method of identification relies on known physical data for the decedent, including personal recollections of family and acquaintances, and ideally, official documentation of sex, age, stature, ancestry and individualized characteristics such as medical history, fingerprints and retained-tissue samples (Warren, 1978). This information is described by Warren as a, "constellation of data" and comprises the body of scientific knowledge about an individual's personal identity.

Idiosyncratic biochemistry and morphology

Idiosyncratic traits are unique to a specific person. For example, every person, with the exception of monozygotic twins, has a unique deoxyribonucleic acid (DNA) pattern. As detailed below, well-preserved bone specimens will yield DNA for comparison with exemplars and/or biologically related family. Similarly, all anatomical structures vary in size and morphology between individuals. Forensic investigators have learned to hone in on specific areas of the skeleton that tend to exhibit the greatest, and most easily discernable, morphological variation. An important forensic issue is that this unique morphology must have been recorded during life. Some of these anatomical variants are a matter of family record (such as extra teeth or a missing finger), while other features are recorded as medical or dental history. Both can be important

Bare Bones: A Survey of Forensic Anthropology

sources of information in a forensic investigation, but it is generally the recorded medical and dental information that proves most valuable in establishing personal identity.

Life-history variables

Life history markers are the evidence of an individual's life that are recorded in the body and include trauma, disease, cultural modifications for medical, surgical or cosmetic reasons, and skeletal indicators of nutritional and environmental stress **(Figure 1)**. Many of these indicators are preserved in the body's hard tissues. Among the contemporary populations of most economically developed countries, evidence of poor nutrition and/or untreated pathology are suggestive of low socio-economic status. Other indicators of low social status might include non-allopathic medical or dental treatment indicative of folk medicine (*e.g.* amateur trephination, or lay dental alteration). Because each person's life history is unique, these cultural and biological indicators serve as excellent sources for evidence of individual identity.

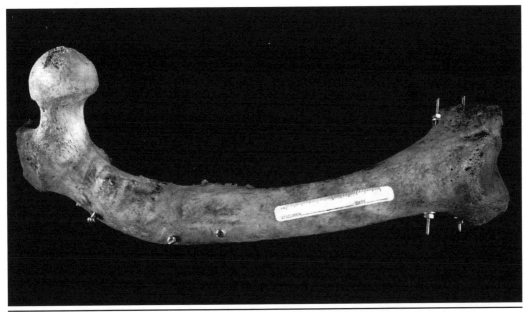

■FIGURE 1: *Surgical femur*

These unique biological traits and life-history markers can be compared to the observable characteristics of an unknown skeleton in order to establish whether the remains are those of a specific individual. This is usually accomplished in one of the following ways: (1) comparison of gross features of the skeleton with known information about the missing person; (2) comparison of antemortem radiographs with postmortem radiographs of the skeleton; (3) comparison of numbers recorded on surgical appliances with antemortem surgical records for the missing person; (4) comparison of the DNA sequence obtained from the bones or teeth of the skeletal specimen with known DNA exemplars or biologically-related family members of the missing person; (5) superimposition of a photograph or radiograph of the missing person over the appropriate specimen; (6) two- or three-dimensional artistic renderings based on skeletal morphology to produce a likeness of the missing person; and (7) analyses of various retained tissues, such as hair, to assist in development of a biological profile of the decedent. Superimposition and forensic art are discussed in a separate chapter.

Comparison of gross features of the skeleton with known information about the missing person

These general features include both the biological profile of group characteristics, as well as gross identifying features known to family and others. These identifying features are often dental anomalies or body modifications. Both group and gross general features are considered presumptive and are usually considered with more specific criteria when establishing personal identity.

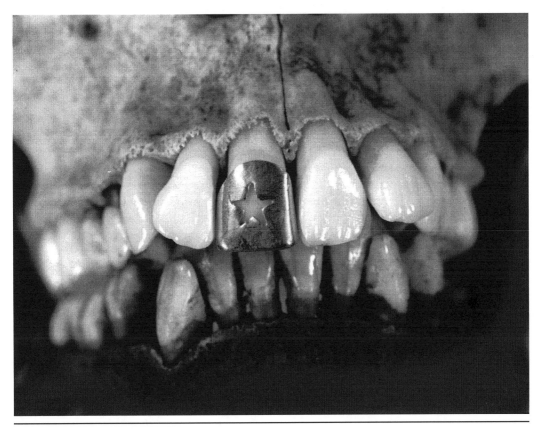

■FIGURE 2: *Starred tooth*

Antemortem/postmortem radiographic comparison

Radiography has served as a primary tool for establishing personal identity from human skeletal remains because it allows investigators to compare clinical radiographs of an individual with postmortem radiographs of skeletal remains. Anthropologists, pathologists and forensic radiologists engaged in human identification target morphological variation, areas of known antemortem disease or pathology, and medical/dental artifacts for comparison with postmortem films (dental comparison will be covered in another chapter on odontology). In addition, radiography has become increasingly important in investigations of mass fatalities, terrorism and war crimes, where it is also used as a screening device for unexploded ordnance, foreign bodies, artifacts, and juvenile remains among commingled adults (Warren *et al.*, 2000; Lichtenstein and Madewell, 1982).

Comparison of antemortem versus postmortem radiographs can be accomplished by gross evaluation, overlaying films, or by overlaying conventional or digital photographs of the films. This can be done via computer or video superimposition. Digitalization enables the investigator to enhance the quality of the antemortem film; or to produce digital pattern tracings to facilitate comparison (Kirk *et al.*, 2002). New strides in diagnostic imaging, most notably computerized tomography (CT) and magnetic resonance imaging (MRI) have also lead to concomitant use of these techniques in personal identification. One advantage to this high-tech approach is the ability to quantify results (Reichs, 1993; Riepert, *et al.*, 1995). Few studies have dealt with quantification as of yet, though some are experimenting with fractal or Fourier analysis to quantify trabecular patterns and gross morphology (*e.g.*, Yu *et al.*, 2003; Majumdar *et al.*, 2000; Parkinson and Fazzalari, 2000).

An additional bonus of CT and MRI is the ability to evaluate the morphology of incidental, non-targeted structures on scans used for diagnosis of trauma or pathology of other anatomical structures of interest, whereas plain film radiographs may show only specifically targeted structures. Future use of these newer technologies will be by necessity, as newer imaging is being used more and more in the clinical setting. For example, plain-film radiography is no longer indicated for evaluation of head trauma in most settings where CT or MRI is available. This results in fewer plain-film skull series and facial views and the loss of several diagnostic features commonly employed in plain-film antemortem - postmortem radiographic comparison, such as frontal sinus patterns and sella turcica morphology.

Cranial features

One area of extreme variation is found in the frontal sinuses, which develop within the frontal bone beneath the supraorbital tori in approximately 95% of all humans (Brogdon, 1998). The frontal sinuses begin development in early childhood and reach their adult size and conformation by age 20. The sinuses are lobulated and divided by a septum, producing a posterior-anterior radiographic pattern reminiscent of a cauliflower floret. This pattern is thought to be unique to each individual, including monozygotic twins (Quatrehomme *et al.*, 1996), and has been described by Harris and colleagues (1987) as a forensic fingerprint.

The frontal sinus pattern remains stable during adulthood, with the exception of slightly increased pneumatization secondary to atrophic changes (Ubelaker, 1984). Pathological changes seen in antemortem films (the reason that sinus films are taken in the first place) enhance, rather than impair, the investigator's ability to match the antemortem pattern with postmortem films of frontal sinuses **(Figures 3a and 3b)**. Reichs (1993) was among the first to use computed tomography for comparison of frontal sinus patterns.

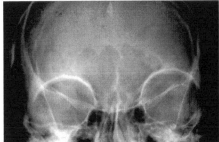

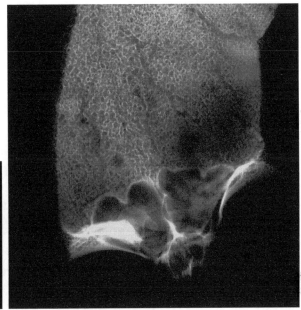

■FIGURE 3A AND 3B: *Frontal sinuses; note the similarities between the antemortem radiograph on the left and the postmortem radiograph on the right. This is the same individual.*

Other cranial features that can be used for antemortem-postmortem comparison include the mastoid sinuses, or air cells (Rhine and Sperry, 1991); the endocranial morphology of the grooves for the middle meningeal artery (Rhine and Sperry, 1991; Messmer and Fierro, 1986); and the shape of the sella turcica in the lateral view (Jablonski and Shum, 1989).

Postcranial features

Vertebral, rib and pelvic features are useful identifiers, particularly because of the frequency with which radiographic views of the chest, and to a lesser degree, the abdomen, are imaged. Almost everyone who has had surgery requiring general anesthesia, or that has been bed-bound in a hospital for a period of several days, has had a chest film recorded. The chest film records 12th rib morphology, the presence of cervical or lumbar ribs, transitional vertebrae, and sternal foramina. Twelfth ribs are highly variable and should be one of the first elements examined for comparison to antemortem radiographs. If the skeletal specimen has relatively short 12th ribs and the antemortem radiograph shows significantly longer 12th ribs, then that missing person can be quickly excluded as a positive match. The same is true for variation in rib number. Vestigial ribs in the cervical or lumbar segments can quickly confirm or exclude specific missing persons as decedents. Transitional vertebrae are vertebrae that exhibit the morphological characteristics that are diagnostic of the vertebrae of adjacent vertebral sections, such as the "sacralized"

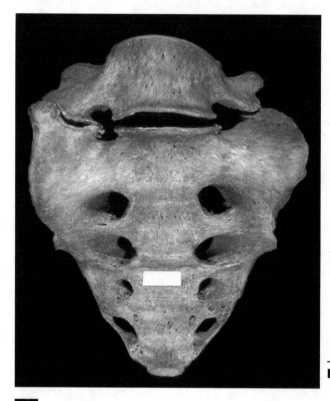

■FIGURE 4: *Sacralized lumbar*

lumbar vertebrae shown in **Figure 4.** Transitional vertebrae are quite common and readily visible on antemortem radiographs. Morphology of the transverse processes, pedicle cross-sections and spinous processes are highly variable and can often be used to quickly exclude an individual as a positive match (Kahana *et al.*, 2002; Valenzuela, 1997; Owsley and Mann, 1992). Abdominal and KUB (kidneys, ureter and bladder) films are also helpful for examining antemortem transverse process shape and size.

Many bony landmarks remain visible over long periods of time. The stability of trabecular bone is related to bone turnover rates, which depend on the location of the landmark and functional aspects of the area. Does a load pass through it? Is it likely to remodel? Some bony disease processes are non-aggressive and slow growing, so can be used for comparison to antemortem films taken many years prior to death **(Figures 5a and 5b)** (Brogdon, 1998; Gold, 2004).

Radiographs taken to visualize soft-tissue structures, such as chest and abdominal films, are often incorrectly exposed for providing good bone detail. Moreover, the antemortem films used for comparison with a postmortem radiograph may be many years old and consequently have lost much of their resolution and contrast. Brogdon (1998) and Fitzpatrick *et al.* (1996) offer several suggestions about how to enhance poor-quality films. The newest x-ray technology utilizes a "digital plate" in lieu of radiologic film., which carries both advantages and disadvantages. A primary advantage is the ability to produce radiographic images without the need for a processor, film and darkroom chemicals. This significantly reduces the previously required storage space for this equipment, as well as archival storage space for exposed film. Digitization of radiographs also enables one to enhance the images via exposure and contrast controls. Digital radiographs are easily shared in consultation with colleagues at a distance in real time. Disadvantages of digital radiography in the forensic arena are the same as digital forensic photography in general – the medium introduces the *possibility* that the image has been manipulated in some way as to make it inadmissible in court (Clement *et al.*, 1995).

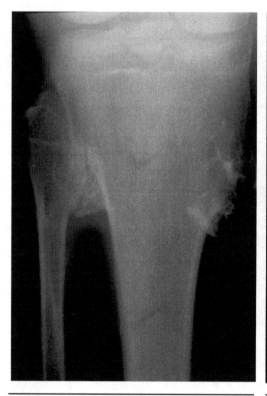 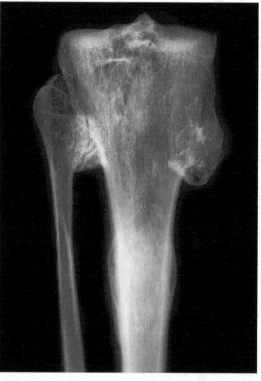

■FIGURE 5A: *An antemortem film of a proximal tibia showing benign tumors and an acute fracture.*

■FIGURE 5B: *Postmortem films showing the same tumors and a healed fracture in the location of the acute fracture seen at left.*

Only practitioners who are thoroughly familiar with normal radiographic anatomy will avoid the pitfalls surrounding the interpretation of comparisons. Many structures appear almost identical between individuals, especially those areas with a strong functional component. Moreover, the manipulation of skeletal material in order to approximate the position of the bones in the antemortem film takes experience. At the very least, advanced graduate and medical students should rotate through a course in musculoskeletal radiology at the university's medical school.

If the anthropologist or pathologist has little experience in examining radiographs for identification purposes, then a clinical or forensic radiologist should be consulted. Apart from antemortem-postmortem comparisons,

Bare Bones: A Survey of Forensic Anthropology

radiologists are particularly helpful with differential diagnosis of rare and unusual bone lesions. Most clinicians are more than willing to view pathological dry-bone specimens because they give a perspective that clinicians do not normally encounter.

Medical and surgical artifacts

In 1993, the Safe Medical Devices Act was passed, requiring all surgeons and other healthcare professionals to record the lot and serial numbers of several classes of orthopaedic devices on the surgical record of each patient. Identification of the vendor, along with the casting, lot and serial numbers on an orthopaedic device, enables the investigator to contact the manufacturer and learn the name of the patient, as well as the surgeon and hospital in which the surgery occurred (**Table 1**)(Ubelaker and Jacobs, 1995). The vendor can be identified by an engraved, stamped or etched logo (**Figure 6; Figure 7**). Some smaller classes of devices, including most of the plates used for long-bone fracture repair, also have serial and lot numbers stamped or engraved on their surfaces. The vendor is often able to provide the investigator with a short list of providers to whom the lot was sold. In instances where the vendor information is not legible, postmortem radiographic images of orthopaedic appliances, plates and screws are also valuable for comparison with clinical films. Investigators can note the position and location of the hardware, count the number of screws and threads, and measure the length of plates and screws for comparison with antemortem films (**Figures 8a and 8b**).

TABLE 1: *Identification via orthopaedic device. Manufacturers of orthopaedic hardware and identifying logos etched or stamped into the hardware. Revised and updated from Ubelaker and Jacobs, 1995.*

Vendor	Identifier
DePuy Orthopaedics	D followed by number
DePuy Ace	For trauma devices; See logo "a" below
Biomet	B; BIOMET; BMT
Biopro	BIOPRO, or logo "h" below
Centerpulse	COUS, or logo below
Intermedics	IOI or logo "d" below
Johnson & Johnson	"J & J"
Ortho Development Corp	ODEV
Orthomet, Inc.	ORTHOMET
Orthopaedic Equipment Co.	OEC; OEC in a triangle
Osteonics	See logo "e" below
Stryker Howmedica Osteonics	See logos "b" and "c" below
Smith and Nephew	n/a
Sulzer Medica	SULZER-MEDICA; SULZER; SOUS
Synthes - Stratec (USA)	See logo "f" below
Wright Medical Tech., Inc.	WMT, plus OM; ORTHOMET
Zimmer	See logo "g" below

Some manufacturers of orthopaedic hardware have either been sold or incorporated into other companies. This information can be valuable in helping determine the time since death. For example, a device with the Intermedics logo indicates a time since death of before *c.* 1995 – the time when the manufacturer stopped using that specific logo, plus additional time for the stock of devices to be sold and implanted.

The following timeline details the periods for several orthopaedic manufacturers that are now under the Zimmer brand. From 1981 until 1997, orthopaedic devices were manufactured under the Intermedics Orthopedics name. In November of 1997, Intermedics became part of Sulzer Orthopedics, Inc. In February of 2000, Sulzer became Centerpulse Orthopedics, Inc., which transitioned to Zimmer in the fall of 2003. This timeline can provide an

Bare Bones: A Survey of Forensic Anthropology

investigator with valuable information about when a device was manufactured and surgically implanted in a living patient. Representatives of various orthopaedic manufacturers are invaluable resources for the history and identifiers of their respective companies. Information about other medical devices is similarly recorded. For example, the website for the North American Society of Pacing and Electrophysiology Heart Rhythm Society's provides information for tracking implanted cardiac pacemakers (http://www.naspe.org/ep-history/timeline).

a. DePuy Ace   b. Stryker-Howmedica   c. Stryker-Howmedica   d. Intermedics

e. Osteonics   f. Synthes-Stratec   g. Zimmer   h. Biopro

■FIGURE 6: *Ortho logos*

FIGURE 7: *Tibial prosthetic with numbers*

Bare Bones: A Survey of Forensic Anthropology

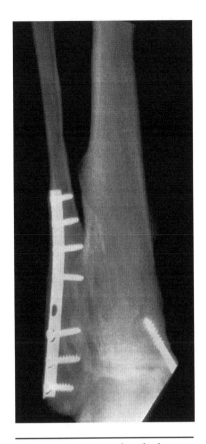

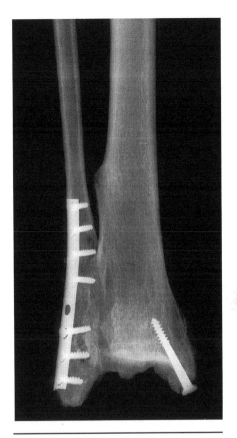

■FIGURE 8A: *AM distal tibia*

■FIGURE 8B: *PM distal tibia*

DNA *from bones and teeth*

Extraction and amplification of nuclear or mitochondrial DNA from bones and teeth provides positive identification of a decedent (e.g., Bender *et al.*, 2000). DNA fingerprinting works hand in hand with morphological assessment by using the information provided via skeletal analysis to produce a short list of possible decedents with whom to compare DNA sequences. The investigation can then proceed to proving the victim's identity through DNA extraction, amplification and sequence analysis. The results of analysis include biostatistical calculations of likelihood of kinship and personal identity (Kemkes-Grottenthaler, 2001).

Success in extracting and amplifying DNA from hard tissues depends on the state of preservation of remains. Bones and teeth that are well-preserved obviously increase the chances of producing amplifiable, and less contaminated, samples of DNA than specimens that have been significantly altered by taphonomic processes. Bones that have become fossilized, cremated, or are otherwise devoid of any fraction of organic component often fail to yield usable DNA (Cattaneo *et al.*, 1999). Perhaps the best bone specimen for DNA analysis is an approximately 3 by 10 cm section of femoral cortex taken from the anterior aspect of the bone. Robust cortical bone is preferred over small, light bones because of the greater density and amount of material (Mulligan, 2005). Teeth are less desirable as specimens because they are more likely to have become contaminated from microorganisms in the surrounding environment (Mulligan, 2005). However, in human-rights work and mass fatalities, it is common to obtain specimens from both bones and teeth.

Apart from positive identification, DNA can be used to establish biological sex and, in some cases, ancestral affiliation of the decedent. This is particularly useful for skeletal material in which the diagnostic elements for sex and ancestry determination are altered or absent. Determination of sex can be accomplished from a sample as small as 1 g of bone (Mannucci *et al.*, 1994; Yamamoto et al., 1998). Determination of geographic ancestry, or "race" from DNA is possible by comparing the profile or haplotype to reference samples to see with which genetic group it most closely aligns (*e.g.*, Allard *et al.*, 2002; Frégeau et al., 1998). Like anthropologists examining various frequencies of morphological traits among and between populations, geneticists are working to determine the characters that define each haplogroup. Haplogroup identifications and frequencies are compared among researchers so unidentified specimens can usually be placed within a broadly defined "racial" group.

At some point in the future, the complete mapping of the human genome may permit investigators to determine eye and hair color, ear-lobe conformation, and other phenotype information from DNA samples collected from skeletons. In underdeveloped areas of the world where medical and dental care are

Bare Bones: A Survey of Forensic Anthropology

not commonly available, DNA may be the only way to determine the identity of the victims. The human-rights arena, as well as the large-scale attempt at genetic identification following the September 11th, 2001 attacks on the World Trade Center in the United States, will prove seminal in exploring the possibilities of handling and processing large numbers of specimens and appropriate databases.

Analysis of other tissues

It is appropriate to mention analysis of human hair in this chapter, as it is relatively resistant to decomposition and is often found at the scene where skeletonized remains are discovered. Time and decomposition often causes hair to lose much of its pigmentation, producing a red tint. However, in many cases it is possible to discern the decedent's hair color and length, which provides valuable general information about the victim's general description for investigators (and as we have seen above in the sections on forensic art). Because the scale pattern of human hair differs from non-humans, analysts can easily weigh in on the question of human versus non-human remains. Gross and microscopic examination of hair can also provide phenotype information used to estimate ancestry. Hair and fiber analysts can also compare an exemplar with unidentified hairs to see if there is consistent morphology and geneticists can extract and amplify DNA from an intact hair root. Hair is one of the sources of latent exemplars most requested by DNA laboratories (Fierro, 1993).

CONCLUSIONS

This chapter briefly summarizes several of the most common methods of establishing personal identity from the human skeleton. A following chapter will discuss the role of forensic art in helping to identify unknown decedents. Human identification is a challenging field in which methods and conclusions must withstand the tests of both scientific peer-review and legal scrutiny. Pathologists and anthropologists tasked with this mission place their careers and reputations on the line with each identification.

References

Allard MW, Miller K, Wilson M, Monson K, Budowle B (2002) Characterization of the Caucasian haplogroups present in the SWGDAM forensic mtDNA dataset for 1771 human control region sequences. Scientific Working Group on DNA Analysis Methods. *Journal of Forensic Sciences* 47(6):1215–1223.

Anderson B (2007) Comment on: Statistical basis for positive identification in forensic anthropology. *American Journal of Physical Anthropology* 133(1):741.

Bender K, Schneider PM, Rittner C (2000) Application of mtDNA sequence analysis in forensic casework for the identification of human remains. *Forensic Science International* 113(1–3):103–107.

Brogdon BG (1998) Radiological identification of individual remains. In Brogdon BG (ed.). *Forensic Radiology*. Boca Raton: CRC Press, Inc., pp. 149–187.

Cattaneo C, DiMartino S, Scali S, Craig OE, Grandi M, Sokol RJ (1999) Determining the human origin of fragments of burnt bone: a comparative study of histological, immunological and DNA techniques. *Forensic Science International* 102(2–3):181–191.

Clement JG, Officer RA, Sutherland RD (1995) The use of digital image processing in forensic odontology: When does 'image enhancement' become 'tampering with evidence'? In Jacob B, Bonte W. (eds.) *Advances in forensic sciences*, Volume 7, Forensic Odontology and Anthropology. Berlin: Verlag Dr Köster, 149–153.

Fierro MF (1993) Identification of human remains. In Spitz WU (ed.) *Spitz and Fisher's Medicolegal Investigation of Death: Guidelines for the Application of Pathology to Crime Investigation*, 3rd Edition. Springfield, Ill.: Charles C. Thomas, pp. 71–117.

Fitzpatrick JJ, Shook DR, Kaufman BL, Wu SJ, Kirschner RJ, MacMahon H, Levine LJ, Maples W, Charletta D (1996) Optical and digital techniques for enhancing radiographic anatomy for identification of human remains. *Journal of Forensic Sciences* 41(6):947–959.

Frégeau CJ, Tan-Siew WF, Yap KH, Carmody GR, Chow ST, Fourney RM. 1998. Population genetic characteristics of the STR Loci D21S11 and FGA in eight diverse human populations. *Human Biology* 70(5):813–844.

Gold M (2004) Hereditary Multiple Exostoses: An identifying pathology. *Proceedings of the American Academy of Forensic Sciences* 10:304.

Harris AMP, Wood RE, Nortje CJ, Thomas CJ (1987) The frontal sinus: a forensic fingerprint?—a pilot study. *Journal of Forensic Odonto-stomatology* 5(1):9–15.

Jablonski NG, Shum BS (1989) Identification of unknown human remains by comparison of antemortem and postmortem radiographs. *Forensic Science International* 42(3):221–230.

Kahana T, Goldin L, Hiss J (2002) Personal identification based on radiographic vertebral features. *American Journal of Forensic Medicine and Pathology* 23(1):36–41.

Kemkes-Grottenthaler A (2001) The reliability of forensic osteology–a case in point. Case study. *Forensic Science International* 117(1–2):65–72.

Kirk NJ, Wood RE, Goldstein M (2002) Skeletal identification using the frontal sinus region: a retrospective study of 39 cases. *Journal of Forensic Sciences* 47(2):318–23.

Lichtenstein JE, Madewell JE (1982) Role of radiology in the study and identification of casualty victims. *Radiologe* 22(8):352–357.

Majumdar S, Link TM, Millard J, Lin JC. Augat P, Newitt D, Lane N, Genant HK (2000) *In vivo* assessment of trabecular bone structure using fractal analysis of distal radius radiographs. *Medical Physics* 27(11):2594–2599.

Mannucci A, Sullivan KM, Ivanov PL, Gill P (1994) Forensic application of a rapid and quantitative DNA sex test by amplification of the X-Y homologous gene amelogenin. *International Journal of Legal Medicine* 106(4):190–193.

Messmer JM, Fierro MF (1986) Personal identification by radiographic comparison of vascular groove patterns of the calvarium. *American Journal of Forensic Medicine and* 7(2):159–162.

Mulligan C (2005) Isolation and analysis of DNA from Archeological, Clinical, and Natural History specimens. In EA Zimmer and E Roalson, (eds.) Methods in Enzymology, Molecular Evolution: Producing the Biochemical Data, Part B. Volume 395. Elsevier Academic Press.

Owsley DW and Mann RW (1992) Positive personal identity of skeletonized human remains using abdominal and pelvic radiographs. *Journal of Forensic Sciences* 37(1): 332–336.

Parkinson IH, Fazzalari NL (2000) Methodological principles for fractal analysis of trabecular bone. *Journal of Microscopy* 198(II):134–142.

Quatrehomme G, Fronty P, Sapanet M, Grevin G, Bailet P, Ollier A (1996) Identification by frontal sinus pattern in forensic anthropology. *Forensic Science International* 83:147–153.

Reichs KJ (1993) Quantified comparison of frontal sinus patterns by means of computed tomography. *Forensic Science International* 61(2): 141–168.

Rhine S and Sperry K (1991) Radiographic identification by mastoid sinus and arterial pattern. *Journal of Forensic Sciences* 36(1):272–279.

Riepert T, Rittner C, Ulmcke D, Ogbuihi S, Schweden F (1995) Identification of an unknown corpse by means of computed tomography (CT) of the lumbar spine. *Journal of Forensic Sciences* 40(1):126–7.

Steadman DW, Adams BJ, Konigsberg LW (2006) Statistical basis for positive identification in forensic anthropology: Response to Anderson. *American Journal of Physical Anthropology* 133(1):741–742.

Steadman DW, Adams BJ, Konigsberg LW (2006) Statistical basis for positive identification in forensic anthropology. *American Journal of Physical Anthropology* 131(1):15–26.

Ubelaker DH, Jacobs CH (1995) Identification of orthopedic device manufacturer. *Journal of Forensic Sciences* 40(2):168–170.

Ubelaker DH (1984) Positive identification from radiograph comparison of frontal sinus patterns. In Rathbun TA and Buikstra J (eds.) *Human Identification: Case studies in forensic anthropology*. Springfield, Ill.: Charles C. Thomas, pp. 399–411.

Valenzuela A (1997) Radiographic comparison of the lumbar spine for positive identification of human remains. A case report. *American Journal of Forensic Medicine and Pathology* 18(2): 215–217.

Warren CP (1978) Personal identification of human remains: An overview. *Journal of Forensic Sciences* 23(2):388–395.

Warren MW, Smith KR, Stubblefield PR, Martin SS, Walsh-Haney HA (2000) Use of radiographic atlases in a mass fatality. *Journal of Forensic Sciences* 45(2):467–470.

Yamamoto T, Uchihi R, Kojima T, Nozawa H, Huang XL, Tamaki K, Katsumata Y (1998) Maternal identification from skeletal remains of an infant kept by the alleged mother for 16 years with DNA typing. *Journal of Forensic Sciences* 43(3):701–705.

Yu JC, Wright RL, Williamson MA, Braselton JP III, Abell ML (2003) A fractal analysis of human cranial sutures. *Cleft Palate-Craniofacial Journal* 40(4):409–415.

Sample Test Questions

1. **Which of the following is the most accurate method of determination of the identity of skeletal remains?**

    a. A drivers license found in the pocket of jeans found in association with the remains

    b. Positive comparison of antemortem and postmortem radiographs

    c. A video superimposition in which all features are consistent

    d. Group characteristics consistent with those of the suspected decedent

2. **Individualizing characters that may be used to determine identity from the skeleton include:**

    a. Orthopedic devices

    b. Frontal sinus pattern

    c. A healed fracture

    d. All of the above

3. **In general, what must exist before a forensic anthropologist can use antemortem/ postmortem radiographic comparison to establish identity?**

    a. Postmortem radiographs and some type of skeletal anomaly

    b. Antemortem radiographs and unique medical/dental treatment or skeletal morphology

    c. A history of trauma in the decedent

    d. Antemortem hospital x-rays and expressed consent of family members

4. **Most anthropologists would consider matching antemortem and postmortem radiographs to be:**

   a. A positive line of evidence for identity

   b. A presumptive line of evidence for identity

   c. An assumptive line of evidence for identity

   d. A poor method for determining identity

5. **The law that allows an anthropologist to trace the serial number on an orthopaedic appliance to the decedent is called the:**

   a. The Orthopaedic Appliances Act of 1898

   b. The Native American Graves Protection and Repatriation Act

   c. FL 411 Medical Examiners Code

   d. The Safe Medical Devices Act of 1993

6. **Fingerprints and DNA are considered:**

   a. Presumptive lines of evidence for identity

   b. Positive lines of evidence for identity

   c. Tentative lines of evidence for identity

   d. Preliminary lines of evidence for identity

Forensic Art

Forensic art is a fascinating blend of art and science. A forensic artist must have a thorough knowledge of human anatomy, an understanding of the scientific research related to the field, and the creative and artistic talent to make the sculpture, drawing or image as life-like as possible.

The final depiction is not expected to be a perfect likeness of the decedent, but rather an approximation of the physiognomy that is close enough so that a family member, friend or acquaintance might make a connection between the image and a missing person. A photograph of the drawing or sculpture is generally placed into a newspaper article that provides additional information about time, location, and other facts surrounding the person's disappearance. This technique has been quite successful in generating leads in investigations that result in the identification of the missing person or decedent.

Forensic art is usually one of the last techniques employed to help determine the identity of the decedent. First, the artist must know the biological profile of the decedent, which is provided by the anthropologist. Secondly, the artist must use either the actual skull of the decedent, or a cast made from the skull. This cannot be done until all other analyses have been performed.

VIDEO SUPERIMPOSITION

Video superimposition utilizes cameras and an electronic video mixing board to enable investigators to superimpose one video image over a second. The technique is most commonly used to overlay a photograph onto a skull, but it can also be used to superimpose a morphological feature onto a radiograph, an antemortem radiograph onto a postmortem radiograph, or even a surveillance photograph onto a photograph of a living suspect (Austin-Smith and Maples, 1992; Ubelaker *et al.*, 1992).

There are two schools of thought in the practice of video superimposition. The first utilizes simple video cameras and an electronic video mixing board with which to overlay images, while the second uses a more high-tech approach, introducing computers and digital-imaging equipment to the former technique. Both methods allow the use of relevant data on skin-tissue thickness; however, the latter permits inclusion of data for facial-growth parameters and degree of fit.

The simpler technique permits the investigator to manipulate the size of the skull image and comparison photograph, but does not permit further alteration of the images. This pioneering phase of video superimposition predated the availability of today's sophisticated computer-imaging software. It is still used by some practitioners who favor the approach because it allows the investigator to answer the inevitable question from the defense attorney as to the possibility of manipulating the image to make it fit (Austin, 1999). Since the answer to this question is "no," a defense counsel and jury cannot suspect that the investigator's work has been influenced by *a priori* knowledge or opinion **(Figure 1)**.

Bare Bones: A Survey of Forensic Anthropology

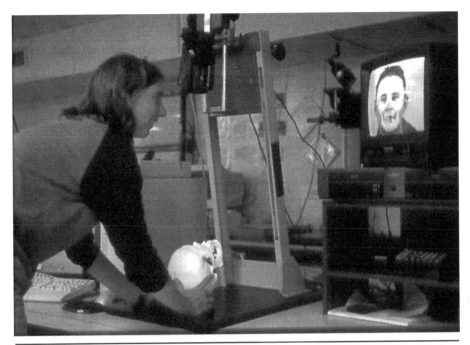

■FIGURE 1: *Anthropologist Laurel Freas positions a skull during a video-superimposition.*

The second school, and the one currently favored by most practitioners, utilizes computers to enhance and manipulate the images in order to reach a decision. Computer programs allow the investigator to improve the sharpness, contrast, and detail of the antemortem photograph. Additionally, these newer methods allow researchers to minimize scaling problems, correct for changes in pitch and rotation of the skull relative to the photograph, and quantitatively evaluate the fit of the superimposition with the aid of high-powered statistical analyses – exacting and highly technical approaches to what was previously an artistic endeavor (Chai *et al.*, 1989; Nickerson *et al.*, 1991' Pesce-Delfino *et al.*, 1986). Even more sophisticated technology is on the horizon, most likely developed in concert with facial recognition software designed with national security in mind.

The following procedure for skull-to-photograph comparison is described by Austin and Maples (1994). The skull is placed under one camera and a photograph is placed under the secondv. The photograph is then focused so that it fills the monitor screen. Tissue thickness markers, if used, are positioned on the skull at prescribed osteometric landmarks. The practitioner then places the skull in the same orientation as the position of the head in the photograph and scales the photograph to the size of the skull. Once the skull and the individual's head in the photograph are comparable in size, the utilized osteometric points are thrown out as a criterion for a favorable match. This task often constitutes the bulk of time spent in producing a video superimposition. Results are evaluated based on agreement between the photograph and several specific anatomic indicators (see Austin-Smith and Maples, 1994 for a representative list of matching criteria). Various methods are used; some are relatively simple, while others are quite elaborate and sophisticated. Accurate results lie not in the method, but in the patience and skill of the investigator. Most investigators agree that experience through trial and error greatly enhances the reliability of the technique. Glassman (2001) has found that the trained "eye" of a forensic artist proves to be, "helpful in both positioning the skull and evaluating video comparisons."

While some courts have ruled that video superimposition constitutes a legal identification, it should be noted that, with the exception of instances of very unique dentition or unusual morphology, most practitioners consider video superimposition as *exclusionary or presumptive evidence*, and not a positive identification.

TWO-DIMENSIONAL FORENSIC ART AND COMPUTER MODELING

Two-dimensional methods, like the three-dimensional techniques discussed below, are a blend of art and science – the exact mixture of which depends on the practitioner. Some employ more art and intuition, while others throw the weight of science at the problem. Two-dimensional art is most often used to identify living suspects based on a witness' recollection of facial features. This can be done the old-fashioned way, or by using special computer programs to produce a composite "photograph" of the decedent (**Figure 2: FACES program**).

Bare Bones: A Survey of Forensic Anthropology

Two-dimensional facial reconstruction is a form of forensic art that seeks to render an accurate representation of a decedent's facial features in the absence of soft tissue. The rendering is based on prior assessment of sex, ancestry, and additional biological parameters of the skull, such as average tissue thickness. Additional clues may include hair, clothing and jewelry found at the scene. If the specimen is from a cold case, an estimation of time since death can be used to render the drawing in appropriate dress or hair style for the period. The art may be produced either freehand or with computer assistance.

■FIGURE 2: *A computer generated composite image, produced by the* Faces 4.0 *program, IQ Biometrix, Inc.*

THREE-DIMENSIONAL FACIAL REPRODUCTION, OR APPROXIMATION

As mentioned above, two-dimensional and three-dimensional forensic-art techniques require prior assessment of the group characteristics, as well as assessment of both biological and cultural clues of idiosyncratic variation. Therefore, it is necessarily one of the last methods used in establishing personal identity. In the past, many reconstructions were relegated to cold cases as a "last resort" when more conventional leads had been exhausted. Rhine (1984) states that three-dimensional facial approximation, or reproduction, is often used as a ". . . means of stimulating investigative leads, with identity being firmly established through the application of other forensic techniques . . .". The methods are not expected to yield a positive identification, but rather to lead the investigators to other sources of identification. Success of the technique seems to rest heavily on the skill of the practitioner, with some artists producing startling results. The technique has been used successfully in some high-profile cases, including the identification of several of the 41 known victims of the "Green River" serial murderer (Haglund and Reay, 1991; Haglund *et al.*, 1987).

Aside from the key factor of artistic skill, the accuracy of facial approximations relies on known, or estimated (based on skeletal analysis), biological parameters, average tissue thicknesses, anatomical knowledge of muscle and gland thickness and conformation, and data about the anatomical relationships of facial features, such as positioning of eyes within the orbits, or the relationship of the fleshy nose to the nasal spine or root (Rathbun, 1984). Early data on skin-tissue parameters was gleaned from cadavers (*e.g.*, His, 1895; Rhine and Moore, 1982; Rhine and Campbell, 1980)(see Table 1). These studies were plagued by the usual associated demographic biases of cadaver-room samples, as well as postmortem differences in skin turgor and muscle tone. Recent work has been performed via sonography, which sample normal *in vivo* tissues and enable inclusion of both healthy adults and children (Manhein *et al.*, 2000). When compared with Rhine and Campbell, as well as much earlier work by His (1895) and Kollmann and Büchly (1898), Manhein and her colleagues found surprising consistency, but with a general trend for greater tissue depth in step with the secular increase in size and better nutrition of contemporary populations (Tables 2 and 3). These data have led to computerized age progression and regression techniques valuable for locating and identifying missing children. Currently, other forms of medical imaging, such as CT and MRI, are being explored as additional sources of tissue thickness data (Stephan, 2002).

TECHNIQUE

Facial approximation begins with careful preparation of either the skull (articulated cranium and mandible) of the decedent, or a cast made from the skull – although Krogman (1987) has outlined a procedure to approximate a missing mandible. Ideally, a cast is used so that the skull can remain under strict evidentiary control; and can be available for examination by other experts.

The forensic artist first applies tissue-depth markers, usually cut from vinyl eraser material. Using available data (see Tables), the artist cuts and places a reference number on each marker that corresponds with Rhine's numbering system for the osteometric points used in reconstruction **(Figure 3)**(Taylor and Gatliff, 2001). The artist then glues the markers at the appropriate landmark. At this point, uniform strips of modeling clay are placed on a flat surface and rolled to the depth of the tissue marker. The artist then applies the strips between adjacent markers. The voids between these strips of clay are then filled in and contoured to the skull and face.

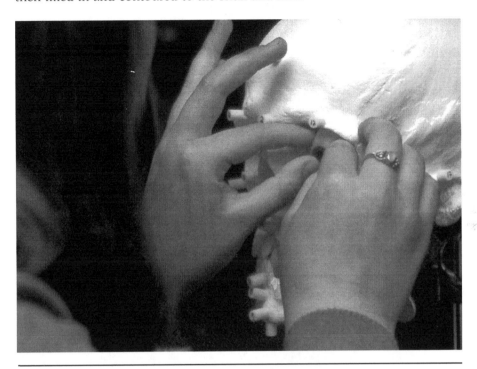

■**FIGURE 3:** *Charlotte Stevens sets the prosthetic eyes into a cast of a skull after placing the skin tissue thickness markers.*

A second approach is the anatomical method, in which the artist/anatomist builds the face by, "sculpting muscles, glands and cartilage on the skull, in effect 'fleshing out the skull'" (Taylor and Gatliff, 2001). Because this method is not based on published normative data, it has not been the single method of choice for forensic presentation. Taylor and Gatliff (2001), arguably the leading forensic artists in the United States, advocate – when circumstances permit – a combination of the two techniques. Keeping step with Glassman's view regarding video superimpositions, both artists are in agreement that the best results are achieved when the approximation is done as a collaborative effort between artist and scientist. Figure 4 illustrates an outstanding approximation using the combination of skin tissue thickness and anatomical techniques. The artist, Angela Canoy, has left the sculpture incomplete to demonstrate the techniques **(Figure 4)**.

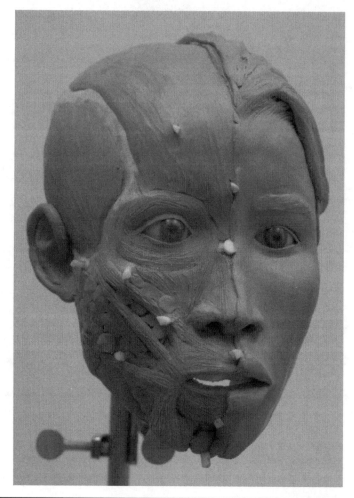

■**FIGURE 4:** *A 3-dimension facial reconstruction by Angela Canoy.*

Some facial approximations bear a startling resemblance to the missing person. While the skill and anatomical expertise of the artist is undoubtedly the primary reason for these successes, most artists will admit that a little luck is helpful as well. If there is no strong evidence of the decedent's weight and robusticity, then the models must be based primarily on published average

Bare Bones: A Survey of Forensic Anthropology

tissue thicknesses. Therefore, in order for the model to closely resemble the decedent, the decedent must have possessed close-to-average tissue thicknesses and anatomical structures with "normal" spatial relationships. Some artists, especially those working in two-dimensional reconstructions, will often produce several images – changing weight, hair styles, eye colors, and other variables to increase the chance of producing a recognizable likeness of the decedent. Fledgling artists interested in honing their skills are encouraged to attend one of the several workshops offered each year by Gatliff, Taylor and other leading artists.

TABLE 1: *Skin tissue thicknesses for adults; Rhine and Moore (1982), His (1895), and Grady et al. (1990)*

| Anatomical Landmark | Rhine & Moore (1982) | | | | His (1895) | | Grady et al. (1990) | | |
| | Blacks | | Whites | | | | | | |
	Male N=N/A	Female N=N/A	Male N=37	Female N=19	Male	Female	Male	Female Thin	Female Well-Nourished
Middle forehead	...	...	...	...	...	...	4.60	2.50	3.60
Glabella	6.25	6.00	5.25	4.75	5.10	4.75	6.60	4.80	6.00
Nasion	6.00	5.25	6.50	5.50	5.55	5.00	6.80	4.90	5.80
Mid nasal bone	...	...	...	...	3.37	3.00	3.30	2.70	2.80
Rhinion	3.75	3.75	3.00	2.75	...	...	3.90	2.50	3.80
Akanthion	12.25	11.25	10.00	8.50	11.49	9.75	12.90	8.90	12.30
Prosthion at upper lip	...	...	...	...	9.51	8.26	11.00	7.30	8.30
Mental sulcus	12.00	...	...	...	10.26	9.75	11.90	10.40	13.20
Mental eminence	11.50	12.50	11.25	10.00	11.43	10.75	9.10	5.50	8.90
Beneath the chin	8.25	8.00	7.25	5.75	6.18	6.50	9.40	5.30	7.50
Mid Supraorbital	4.75	8.00	8.25	7.00	5.89	5.50	8.00	5.40	7.40
Mid Suborbital	7.50	8.25	5.75	6.00	5.08	5.25	11.60	5.60	9.60
Lateral orbit	...	...	...	...	...	...	13.00	7.70	11.70
Mid zygomatic arch	...	...	...	...	...	...	10.30	6.30	10.00
Supraglenoid	...	...	...	...	6.07	6.75	13.30	7.60	11.40
Anterior masseter	...	...	...	...	8.65	8.10	13.50	9.00	12.70
Mid ramus	...	...	...	...	18.05	17.05	24.30	14.30	22.90
Gonion	14.25	13.50	11.50	12.00	12.21	11.50	15.50	8.30	14.60
Anterior zygomatic	13.25	13.00	10.00	10.75	...	...	...	...	...
Sup mid mandible	16.50	17.00	16.00	15.50	...	...	...	...	...
Chin lip fold	11.75	12.25	10.75	9.50	...	...	...	...	...
Posterior maxilla	22.00	20.25	19.50	19.25	...	...	...	...	...

*Note that the anatomical landmarks and nomenclature used by various researchers vary slightly.

TABLE 2: *Manhein et al., 2000. Average adult skin tissue thicknesses as recorded in vivo with ultrasound. All measurements made to nearest 1 mm.*

MANHEIN, LISTI, BARSLEY, MUSSELMAN, BARROW AND UBELAKER, 2000; ADULT IN VIVO MEASUREMENTS OF SKIN THICKNESS

| | BLACKS | | | | | WHITES | | | | | | | |
| | 19–34 YEARS | | 35–45 YEARS | | 46–55 YEARS[1] | 19–34 YEARS | | 35–45 YEARS | | 46–55 YEARS | | >56 YEARS | |
LANDMARKS	MALE N=19	FEMALE N=18	MALE N=3	FEMALE N=21	FEMALE N=5	MALE N=28	FEMALE N=52	MALE N=10	FEMALE N=15	MALE N=5	FEMALE N=6	MALE N=5	FEMALE N=9
Glabella	5.2	4.6	5.3	4.5	4.8	5.0	4.8	5.5	4.7	6.0	4.8	5.6	5.2
Nasion	6.6	6.0	5.7	5.2	6.0	6.0	5.5	6.4	5.3	7.2	6.2	6.6	6.0
Rhinion	2.2	1.7	1.7	1.5	2.0	1.9	1.8	2.4	1.6	1.8	1.8	2.0[+]	1.8
Lateral nostril	9.2	8.4	10.3	8.4	8.4	7.5	8.6	9.8	8.0	10.4	10.8	10.8	9.8
Mid-philtrum	13.0	9.2	11.0	8.8	8.2	11.9	9.1	10.6	7.4	8.03	8.0	9.4	8.0
Chin lip fold	12.7	11.8	12.7	11.7	10.0	11.1[2]	10.3	13.1	9.6	11.6	9.8	12.2	11.4
Mental eminence	12.1	10.8	12.3	11.2	10.8	10.0	9.2	12.0	9.2	11.0	10.7	11.8	12.3
Beneath chin	8.8	6.7	7.0	6.4	7.2	7.2[2]	6.0	8.0	5.4	7.2	6.7	5.6	8.0
Supraorbital	6.4	6.1	6.3	6.0	5.8	5.3	5.7	5.9	5.5	7.7	6.5	5.6	6.3
Suborbital	5.8	6.2	7.0	6.9	5.8	5.8	6.1	6.2	5.7	6.8	7.3	5.0	7.0
Supracanine	12.8	10.0	10.3	9.6	9.0	11.9[2]	9.3	10.1	7.8	10.0[3]	7.7	9.2	8.0
Subcanine	14.4	10.9	10.7	11.5	12.4	11.5	9.4	10.2	8.7	10.0	9.0	11.8	9.7
Posterior maxilla	28.2	26.6	27.3	26.8	26.8	28.5	26.3	24.6	25.1	28.2	27.2	23.6	29.4
Sub mid mandible	24.5	21.7	23.7	22.5	21.2	25.1	23.4	21.1	20.1	21.4	21.7	20.6	27.2
Inf mid mandible	14.1	12.6	13.3	13.1	13.4	14.8	13.7	15.6	12.6	15.4	13.0	11.4	17.4
Lateral eye orbit	4.8	5.0	3.7	4.9	4.8	4.2	4.7	4.3	4.3	5.4	4.5	5.2	4.9
Ant zygoma	8.4	10.2	6.3	9.8	9.8	7.8	9.3	8.2	8.7	8.2	10.2	6.4	11.0
Gonion	21.1	17.0	20.7	16.2	14.8	20.0	17.4	19.6	15.3	19.0	14.7	14.0	16.9
Root of zygoma	7.4	6.4	5.7	5.6	6.0	7.8	7.4	6.6	4.9	5.4	6.0	5.2	7.4

[1] No black male data for 46-55 years or >56 years; no black female data for >56 years.

[2] N=27; excludes men with beards and mustache

[3] N=3; excludes men with beards and mustaches

[4] N=4; excludes men with beards and mustaches

TABLE 3: *Manhein et al., 2000. Average child skin tissue thicknesses as recorded in vivo with ultrasound. All measurements made to nearest 1 mm.*

MANHEIN, LISTI, BARSLEY, MUSSELMAN, BARROW AND UBELAKER, 2000; CHILD IN VIVO MEASUREMENTS OF SKIN THICKNESS

LANDMARKS	AGES 3 TO 8 YEARS						AGES 9 TO 13 YEARS						AGES 14 TO 18 YEARS					
	BLACK		WHITE		HISPANIC		BLACK		WHITE		HISPANIC		BLACK		WHITE		HISPANIC	
	M	F	M	F	M	F	M	F	M	F	M	F	M	F	M	F	M	F
	N=37	N=52	N=36	N=43	N=3	N=6	N=62	N=59	N=45	N=51	N=8	N=9	N=12	N=25	N=27	N=35	N=4	N=1
Glabella	4.1	4.0	4.0	3.9	4.7	4.2	4.5	4.3	4.6³	4.4	4.1	3.8	5.3	4.7	5.0	4.6	4.5	7.0
Nasion	5.4	4.9	5.7	5.0	6.3	5.0	5.4	5.4	5.7³	5.5	4.9	5.3	6.1	5.3	6.3	5.4	4.8	5.0
Rhinion	1.8	1.7	1.8	1.7	1.7	1.7	1.9	1.7	1.6	1.5	1.6	1.6	2.1	1.7	2.0	1.8	1.5	1.0
Lateral nostril	7.3	7.0	7.2	7.0	6.3	6.3	7.4	7.6²	7.4	7.7	7.9	5.7	7.9	8.1	7.8	7.7	5.0	9.0
Mid-philtrum	9.0	8.9	9.0	8.3	7.3	8.0	10.0	9.6	9.7	9.4	9.3	9.2	12.1	9.9	11.2	9.4	11.5	8.0
Chin lip fold	8.6	8.2	8.1	7.6	7.0	8.7	9.8	10.3	9.6	9.0	10.0	9.2	12.6	10.1	10.4	9.7	11.3	11.0
Mental eminence	8.3	8.3	8.3	7.4	6.0	8.0	9.9	10.0	8.7	8.8	8.4	8.4	9.5	10.0	9.3	8.7	10.3	15.0
Beneath chin	4.5	4.8	4.6	4.2	4.7	4.2	5.5	5.8	5.5	5.5	5.1	5.1	6.3	5.6	6.0	5.5	5.8	9.0
Supraorbital	4.5	4.5	4.6	4.4	4.3	4.2	5.2	5.3	5.2	5.1	4.9	4.9	5.8	5.7	5.7	5.7	5.5	7.0
Suborbital	5.6	5.6	5.5	5.6	5.0	5.5	5.8	6.1	5.9	5.6	6.4	5.0	6.0	6.4	5.3	6.0	5.8	10.0
Supracanine	8.9	8.8	9.4	8.4	8.0	9.3	10.7	10.0	10.0	9.8	10.0	10.3	12.3	10.6	11.7	10.3	12.0	11.0
Subcanine	8.5	9.0	8.4	7.9	6.7	8.2	11.0	10.2	9.6	9.2	10.8	8.3	12.8	11.0	10.6	9.8	10.0	10.0
Posterior maxilla	22.1	23.0	23.3	22.7	19.7	24.8	23.6	24.5	24.7	24.3	24.4	24.6	26.0	27.6	27.4	26.8	25.3	32.0
Sub mid mandible	17.4	18.0	20.7	18.9	14.7	20.8	20.1	20.0	21.6	20.8	21.4	20.0	21.9	23.2	23.2	23.2	21.0	24.0
Inf mid mandible	8.7	9.8	10.4	10.5	7.3	11.5	10.3	10.8	12.1	11.7	10.8	11.3	11.2	12.0	12.3	13.4	10.3	18.0
Lateral eye orbit	4.1	3.9	4.1	4.0	3.0	4.3	4.4	4.4	4.4	4.6	4.6	3.8	4.4	4.6	4.3	4.5	4.3	5.0
Ant zygoma	7.8	8.3	8.4	8.4	6.3	8.5	8.3	8.9	9.1	9.5	8.4	7.4	7.3	9.2	8.0	9.5	7.8	14.0
Gonion	12.8	13.5	13.7	13.9	13.7	14.0	14.7	14.6	15.4	14.4	15.4	14.6	17.9	16.2	18.1	17.0	15.3	24.0
Root of zygoma	4.2	4.7	4.8	4.6	4.3	4.3	5.0¹	4.8²	5.4	5.2	6.3	4.6	6.0	6.2	6.0	6.8	4.8	8.0

[1]N=61
[2]N=58
[3]N=44

References

Austin D (1999) Video superimposition at the C.A. Pound Laboratory 1987 to 1992. *Journal of forensic sciences* 44(4):695-9

Austin-Smith DE, Maples WR (1992) Photograph/photograph video superimposition in individual identification of the living. Presented at the 44th Annual Meeting of the American Academy of Forensic Sciences, New Orleans, LA. February 21, 1992.

Austin-Smith DE, Maples WR. 1994. The reliability of skull/photograph superimposition in individual identification. *Journal of Forensic Sciences* 39(2):446-455.

Chai D-S, Lan Y-W, Tao C, Gui R-J, Mu Y-C, Feng J-H, Wang W-D, Zhu J (1989) A study on the standard for forensic anthropological identification of skull-image superimposition. *Journal of Forensic Sciences* 34(6):1343-1356.

Glassman DM (2001) Methods of superimposition. In, Taylor KT, Forensic Art and Illustration. Boca Raton: CRC Press, Inc., pp. 477-498.

Haglund WD, Reay DT (1991) Use of facial approximation techniques in identification of Green River serial murder victims. American Journal of Forensic Medicine and Pathology 12(2):132-142.

Haglund WD, Reay DT, Snow C (1987) Identification of serial homicide victims in the 'Green River Murder' investigation. *Journal of Forensic Sciences* 32(6):1666-1675.

His W (1895) Johann Sebastian Bach's Gebeine und Antlitz nebst Bemerkungen über dessen Bilder. *Abhandlung durch Mathematik und Physik* 22:380-420.

IQ Biometrix, Inc., Faces 4.0.0.0, in collaboration with CogniScience, Inc.

Kollmann J, Büchly W (1898) Die persistenz der rassen und die reconstruction der physiognomie prähistorischer Schädel. *Archiv für Anthropologie* 25:329-359.

Krogman WM (1987) Method for approximation of a missing mandible based on the cranium. In Krogman and Ýþcan (eds.) *The Human Skeleton in Forensic Medicine*. Springfield, Ill.: Charles C. Thomas.

Manhein MH, Listi GA, Barsley RE, Musselman R, Barrow NE, Ubelaker D (2000) *In vivo* facial tissue depth measurements for children and adults. *Journal of Forensic Sciences* 45: 48-60.

Nickerson BA, Fitzhorn PA, Koch SK, Charney M (1991) A methodology for near-optimal computational superimposition of two-dimensional digital facial photographs and three-dimensional cranial surface meshes. *Journal of Forensic Sciences* 36(2):480-500.

Pesce-Delfino V, Colonna M, Vacca E, Potente F, Introna F (1986) Computer-aided skull/face superimposition. *American Journal of Forensic Medicine and Pathology* 7(3):201-12.

Rathbun TA (1984) Personal identification: Facial reproductions. In Rathbun TA and Buikstra JE (eds.) *Human Identification: Case studies in forensic anthropology*. Springfield, Ill.: Charles C. Thomas, pp. 347-356.

Rhine JS (1984) Facial reproduction in court. In Rathbun TA and Buikstra JE (eds.) *Human Identification: Case studies in forensic anthropology*. Springfield, Ill.: Charles C. Thomas, pp. 357-362.

Rhine JS and Campbell HR (1980) Thickness of facial tissues in American Blacks. *Journal of Forensic Sciences* 25(4):847-858.

Rhine JS and Moore CE (1982) Facial reproduction tables of facial tissue thicknesses of American Caucasoids in Forensic Anthropology. Albuquerque: *Maxwell Museum Technical Series*, #1.

Stephan CN (2002) Facial approximation: globe projection guideline falsified by exophthalmometry literature. *Journal of Forensic Sciences* 47(4):730-735.

Taylor KT and Gatliff BP (2001) Skull protection and preparation for reconstruction. In Taylor KT (ed.) *Forensic Art and Illustration*. Boca Raton, Fla.: CRC Press.

Ubelaker DH, Bubniak E, O'Donnell G (1992) Computer-assisted photographic superimposition. *Journal of Forensic Sciences* 37(3):750-762.

Sample Test Questions

1. **An advantage to the FACES program, or computerized facial renderings, is:**

   a. Provides the witness with choices

   b. It is quicker and more efficient than hand-drawn sketches

   c. The result almost always looks more real than hand-drawn sketches

   d. All of the above

2. **Video superimpositions are rarely used in the courtroom because:**

   a. The technique is primarily used to exclude a missing person as the decedent, so when it "matches", it provides only a presumptive line of evidence for identity

   b. The defense usually stipulates the identity of the decedent, so there is no need to introduce identifying evidence

   c. Superimpositions are inflammatory, in that they show the decedent's photograph over a skull

   d. All of the above

3. **The two different approaches to 3D facial approximation (which can also be combined) are:**

   a. Skin tissue thickness, artistic

   b. Anatomical, skin tissue thickness

   c. Anatomical, Boasian

   d. Anatomical, physical

Section III

Interpreting Skeletal Trauma

Anthropologists have not always played a role in evaluating skeletal trauma in forensic contexts. Early practitioners felt that determination of cause of death was best left to the pathologist and the realm of forensic medicine (Krogman, 1962). However, later, T. Dale Stewart (1979) acknowledged that anthropologists might be helpful in determining the cause of death by providing a careful description of bone damage and its relationship to vital anatomical structures and helping pathologists differentiate perimortem trauma from postmortem damage to bone. By the mid-1980s, forensic anthropologists began to play a more important role in the analysis of traumatic injuries to the skeleton (Maples, 1986). Maples and colleagues around the country developed close ties with medical examiners, exchanging knowledge and experience and enhancing the quality of death investigations - particularly those involving decomposed or skeletonized remains. Today, forensic anthropologists can be instrumental in helping the coroner or medical examiner determine the mechanism of traumatic injury and cause of death. In fact, several large medical examiner's offices employ forensic anthologists full-time, often involving them in "fresh" cases as well as their usual duties examining skeletal remains.

The skeleton often records traumatic injury. The ways in which bone responds to traumatic injury are well-documented, and so provide lasting evidence of injury. All people share similar anatomy and are subjected to the same types of external forces. Bones break in predictable ways – a fall onto an outstretched hand, a twist of the ankle – so that most common fractures have been named. These named fractures are each caused by a known mechanism. For example, a fall from a skateboard onto an outstretched hand may produce a *Colles fracture* of the distal radius; or striking an object (or a person) with a closed fist may result in a Boxer's fracture of the 4th or 5th metacarpal. Because these fractures result from known mechanisms, examination of the skeleton can allow the forensic anthropologist to better understand the traumatic event and to provide the pathologist with valuable information about the decedent's life history and the events surrounding their death.

BONE TRAUMA

Trauma is classified as having occurred during either the **antemortem, perimortem** or **postmortem** intervals. It is interesting to note that these intervals differ between pathologists and anthropologists. Pathologists are able to mark differences in tissue reactions between injuries that occurred while the heart was still beating, and tissue damage that occurred immediately after cessation of heart activity by, for example, assessing the presence or absence of bleeding into adjacent tissues. The organs and tissues respond differently once death has occurred. Anthropologists, however, must work within broader intervals, based on evidence of healing, lack of healing, or taphonomic alteration of bones. An example involves cases in which the victim's body has been dismembered after death. If soft tissues are present, the pathologist would recognize whether or not the dismemberment took place after the victim had expired. In the absence of soft tissues, the anthropologist would be presented with bones that demonstrate sharp-force trauma or saw marks which would not show evidence of healing, but also would not have been altered by taphonomic processes. The inevitable question may arise: was the victim alive when the murderer began to dismember the body? Based on skeletal analysis alone, it is impossible to know. Experience dictates that if the saw marks are uniform and neat, then no struggle took place, and it can be assumed that the victim was at least unconscious during the dismemberment; but the anthropologist cannot answer these questions with the certainty of a pathologist, who has

Bare Bones: A Survey of Forensic Anthropology

more information with which to reach his or her conclusions. In the paragraphs that follow, we will discuss each interval and the ways in which it is possible to pinpoint when an injury or bone damage occurred.

Antemortem trauma

Antemortem trauma is an injury that occurred during life and has either healed or is in the process of healing at the time of death. When a bone is fractured, it immediately begins the healing process. Most clinicians describe five distinct phases of bone repair and remodeling (Adams and Hamblen, 1998). Healing begins when the fractured ends of the bone and adjacent periosteal tissue bleed into surrounding fascia and muscle, creating a hematoma. Next, special cells, called osteoblastic precursor cells, begin to proliferate and form a collar of new bone that bridges the gap across the fracture. These precursor cells soon give way to osteoblasts that lay down a matrix of immature, or *woven bone*, called a *callus*. The callus is gradually transformed into typical lamellar bone which continues to strengthen and remodel along lines of stress. This process takes a variable amount of time depending on the severity of the fracture, the site of the fracture (and the adequacy of its nerve and blood supply), the extent to which the fracture site can be immobilized, and the general health of the individual. In most cases, long bone will have reached the remodeling phase by four to six weeks. If a bone has completely healed, it is not possible to determine how long ago it was initially broken.

In adults, alignment of the two fractured ends of a bone are rarely perfect. So, although bone remodels when it has been damaged, fractures and defects that occurred many years prior to death may still leave clues for the attentive anthropologist. If a skeletal examination reveals that the right radius and ulna were previously fractured, then the examiner has at least one line of evidence to compare with the known information about a missing person (**Figure 1**).

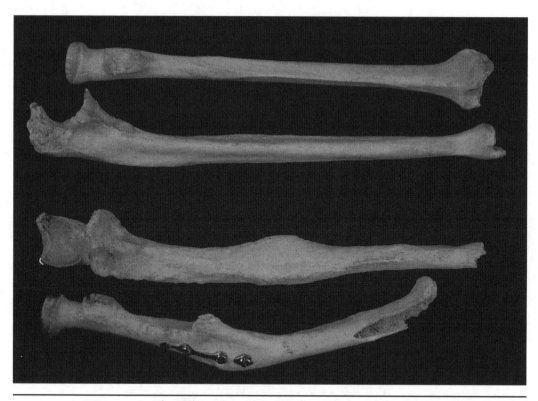

■FIGURE 1: *The right radius and ulna show evidence of antemortem fracture and healing. The radius was repaired by internal fixation with an orthopaedic plate. The distal ends of both bones also show postmortem damage by scavengers.*

Perimortem trauma

Perimortem trauma is an injury that occurred around the time of death and therefore may have contributed to the death. Perimortem trauma shows no evidence of healing or having been subjected to the taphonomic processes discussed in a previous chapter. Perimortem trauma can occur as a result of three mechanisms: blunt forces, sharp forces, or ballistic projectiles (usually gunshot wounds in the United States). Each of these distinct mechanisms is discussed below under *Types of Trauma*.

Bare Bones: A Survey of Forensic Anthropology

Postmortem trauma

Postmortem trauma is defined as damage to the body after death. As discussed in a previous chapter on taphonomy, this damage can be caused during the immediate postmortem interval, or it can occur over a long period of time in the form of weathering, cortical erosion, and postmortem fracturing.

Fractures that occur soon after death can be problematic because they occur during the gray area when the environment or human and animal activity has yet to begin to alter the appearance of the bones. An important clue lies with the general characteristics of the damage. Does the fracture or lesion make sense from a clinical perspective? In other words, is the fracture we are assessing common, or even possible, in living people? Is it a named fracture with a known mechanism? If a fracture is in an unusual location, or exhibits a characteristic that cannot be explained as having occurred through the usual forces exerted in a living, articulated human, then the anthropologist should be suspicious that the lesion is postmortem and occurred when the bone was part of a disarticulated skeleton.

The margins of the fractures also provide clues. If bones are damaged long after death and decomposition has occurred, the bones are often discolored by surrounding soils. If the edges of the fracture are not discolored, then the breakage occurred long after the environment began to exert its taphonomic forces on the bones.

TYPES OF TRAUMA

Trauma may occur through impact against, or by, a blunt object (*e.g.*, a fist, bat or car windshield), by a sharp instrument (*e.g.*, a knife), or by a ballistic projectile. Each of these mechanisms produces different effects on bone.

Blunt-Force Trauma

Blunt-force trauma is caused when a blunt object strikes the body, or when a moving body strikes a blunt object or the ground. In the cranium – usually the most significant site of injury in homicide cases - fractures emanate from the point of impact and radiate outward. These fractures find the path of least resistance, but also propagate along dense, hard bones with a low elastic limit, such as the ridge of the petrous portion of the temporal bone. Experiments have shown that it is difficult to predict fracture patterns produced by a given

Interpreting Skeletal Trauma

mechanism and that the damage observed is often counter-intuitive. For example, rectangular objects can create circular defects composed of concentric fractures (**Figure 2**). As a result, blunt-force injuries are often difficult to interpret, and many anthropologists limit their opinions to general statements concerning mechanism of injury. In some cases, it may be possible to determine the sequence of two or three blows by interpreting radiating fracture patterns. Later fractures will terminate into earlier fractures. Multiple blows, however, may be impossible to interpret because of their complexity. Fractures propagate through bone at an extremely high rate of speed. A single blow may create radiating fractures that terminate into each other. Beyond three or four blows, many anthropologists will use the phrase "multiple blows" in their report.

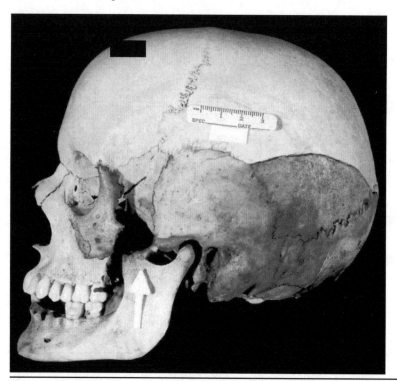

■FIGURE 2: *Blunt-force trauma to a skull. The zygomatic bone has a typical "tripod fracture". The fracture line follows the squamosal suture of the temporal bone and extends across the parietals. The arrow points to a fracture of the coronal process of the mandible. The differing color of the fractured segments show that these fragments fell away during decomposition. This skull has been reconstructed.*

Bare Bones: A Survey of Forensic Anthropology

Tubular long bones fracture in predictable ways. An impact to the shaft of a tubular bone places the bone in compression at the site of the impact and creates tension on the opposing side. Bone will almost always fail in tension before it will fail in compression. The way in which a fracture line propagates is related to the direction of the force relative to the axis of the bone. Fractures produced by a force that is exerted perpendicular to the axis of the bone are termed *transverse* or *oblique* fractures. In some cases the bone begins to fail in tension, but the fracture does not completely transect the bone. This type of *greenstick* fracture is often seen in children. Another type of fracture due to a perpendicular force is called a *butterfly* fracture, which produces a distinct pattern that enables one to easily determine the direction of impact. Torsion, or a twisting force, produces a *spiral* fracture. Compression along the axis of the bone can crush, or comminute the bone, creating a fracture with many fragments.

Fractures to the hyoid bone merit special attention because they are significant in forensic cases. Research has shown that the hyoid bone breaks approximately 40% of the time in cases of known manual strangulation. Of course this means that it fails to fracture the majority of the time. Therefore, the absence of a hyoid fracture does not rule out the possibility that the victim was strangled. On the other hand, manual strangulation was the cause of hyoid fracture in the vast majority of cases with a confirmed hyoid fracture, and so presents strong evidence of foul play (Ubelaker, 1992).

Patterned Injuries

Some defects in bone provide a clue to the type of tool used as the weapon. These wounds are called patterned injuries because they produce a pattern that approximates the margins or outline of the weapon (Clark and Sperry, 1992). **Figure 3** shows a lethal patterned defect in the cranial vault of a store clerk. The pattern matches that produced by a golf club iron (the end of the shaft, or *hosel*, and the bottom edge of the face of the club has produced a depressed skull fracture with two distinctly different surfaces). The golf club was taken from a bag within the store, used during the assault and later found off premises. The club had blood stains that matched the victim's blood type.

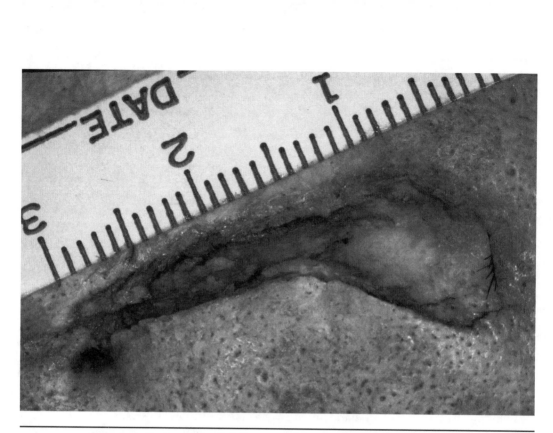

■FIGURE 3: *A patterned defect caused by the shaft and head of a golf club iron.*

Conventional wisdom among criminalists states that offenders usually employ a single weapon during a homicide. When patterned injuries are present that seem to have completely different characteristics, the investigator must consider how a single weapon could have produced both defects (**Figures 4a and 4b**). The late Dr. William R. Maples, a renowned forensic anthropologist, often took his students to the hardware store to see what types of tools were available that might produce a wound seen in a case under investigation at the laboratory. Students would then buy various types of hammers, screwdrivers, or whatever was the class of tool suspected to be similar to the offending weapon, and return to the laboratory to pound on soft spruce wood in an effort to reproduce the defect seen in the bones. Unusually shaped defects pose a special challenge. Law enforcement detectives will often re-visit the scene of a crime and search for likely tools or makeshift weapons. Even when investigators have the suspected murder weapon in hand, they will often wait until the forensic anthropologist reports the class of weapon to see if it corroborates their findings. Then the investigators can ask if it is possible that the weapon in their possession could have been used to commit the crime. This lends added weight to the anthropologist's testimony.

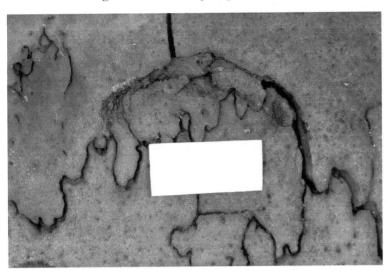

■**FIGURE 4A:** *A hexagonal pattern, approximately one inch in diameter, created by the head end of a claw hammer.*

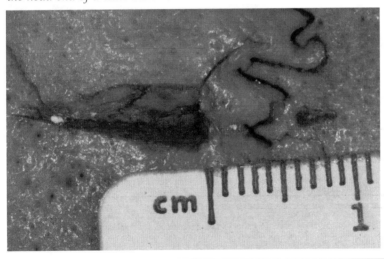

■**FIGURE 4B:** *Two triangular patterns caused by the claw end of the same hammer. The smaller defect is very subtle, but can be seen on the opposite side of the sagittal suture. The indentations differ in size because the blow was at an angle.*

Sharpforce Trauma

Sharpforce trauma is essentially blunt force applied to a very small area. The sharp blade of a knife produces sufficient force over such a small area that the bone is unable to respond and fails at the point of contact. Sharp instruments can produce incisions, puncture/stabbing wounds and chopping/hacking injuries.

Incised wounds produce a V-shaped cross section. Since the knife rarely passes perfectly along its course, one side of the defect is usually lifted relative to the opposite side. The characteristics of incised wounds are dependent on the type of blade, the force applied and the track of the weapon across the tissues. Therefore, a number of different-looking cutmarks could have been made with the same weapon

Stabbing wounds are injuries caused by a sharp-ended weapon being thrust along its long axis into tissues. These types of defects often yield information about the tool characteristics that include the width, depth, and cross-sectional shape of the knife or instrument. For example, an ice pick leaves a small, circular defect; a flat screwdriver leaves a rectangular defect; and the defect from a knife may reveal a blunt upper margin and a sharpened lower margin.

Chopping or hacking trauma, such as that caused by an axe or machete, may leave striae recording irregularities along the length of the blade. Experts in toolmark analysis may be able to lift molds from the bony defect and match them with molds taken along the edge of a blade suspected of being the murder weapon **(Figure 5)**.

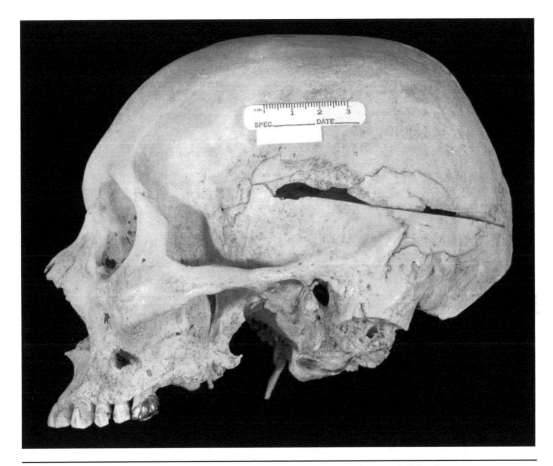

■FIGURE 5: *This cranium exhibits at least three chop marks, probably from a machete or cane cutter.*

Saw Marks in Bone

It is, unfortunately, not uncommon for killers to dismember the bodies of their victims, ostensibly to allow for easier transport or disposal of the body, or to make it more difficult to establish the identity of the victim. Dismemberment is usually accomplished by knives, or saws, or both. If a saw was used, it is possible for investigators to examine the saw marks on bone and establish the characteristics of the saw blade.

As a saw passes through a given material, it creates a *kerf*, a void representing the width of distance between the teeth on the saw blade. Once the material is completely cut into two segments, it is possible to examine the walls of the material through which the blade has passed. This kerf wall is the diagnostic feature used by anthropologists to determine the type and characteristics of the saw **(Figure 6)**. The degree of blade wear impacts the appearance of the kerf wall, but not to the extent that it makes identification of the blade characteristics impossible (Freas, 2006).

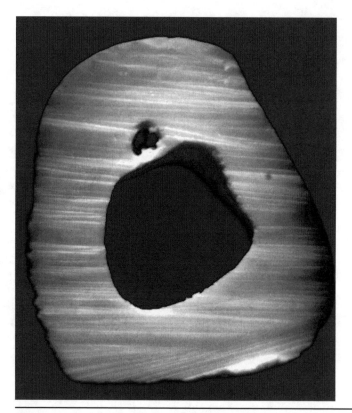

■FIGURE 6: *A kerf wall showing the striae created by the passing teeth of the saw blade. Photograph courtesy of Laurel Freas.*

Quantitative analyses of saw-mark patterns is extremely complex and researchers have achieved only limited success (Freas, 2006; Bartelink *et al.*, 2001). Therefore, evaluation of saw marks remains largely qualitative, with experience playing an important role in correct interpretation. Most anthropologists should limit statements about saw marks to the class characteristics of the saw, such as the type of saw (*e.g.*, crosscut saw, rip saw, hacksaw, pruning saw, etc.), and perhaps comments pertaining to teeth per inch or the "set" of the teeth. Courtroom testimony in this area will address whether it is possible that a specific saw – obtained by investigators – could have produced the defects seen on the bone. This area of expertise is limited to only a few practicing anthropologists in the country, most notably Dr. Steve Symes of Mercyhurst College (see Symes, 1992; Symes *et al.*, 1998).

BALLISTIC TRAUMA

Ballistic trauma includes wounds caused by bullets, arrows, shrapnel and other projectiles that strike tissue. The branch of ballistics relevant to forensic anthropology is called ***terminal ballistics***, which is the study of the behavior of a projectile when it hits its target. A detailed discussion of ballistics and firearms is not appropriate for this volume, but readers can refer to the book by DiMaio (1998) for further information.

Several questions may be answered by examination of ballistic injuries in bone. Two principle questions are the caliber of the projectile and its trajectory. Ballistic defects in bone provide a long-lasting record of gunshot wounds and often provide the investigator with better information about caliber and trajectory than do wounds in soft tissue. Furthermore, bones which show evidence of ballistic injury may be retained as evidence, to be later examined by additional experts or presented in court.

Ballistic wounds can be either penetrating or perforating. ***Penetrating wounds*** enter the target (for our purposes, a person) but do not exit. ***Perforating wounds*** leave both an entry wound and an exit wound as the projectile passes completely through the body (**Figure 7**).

The size, location, number and characteristics of projectile defects in bone may allow the investigator to make statements about the type of weapon used and other factors that have probative value in reconstructing the events leading to death. Among these are:

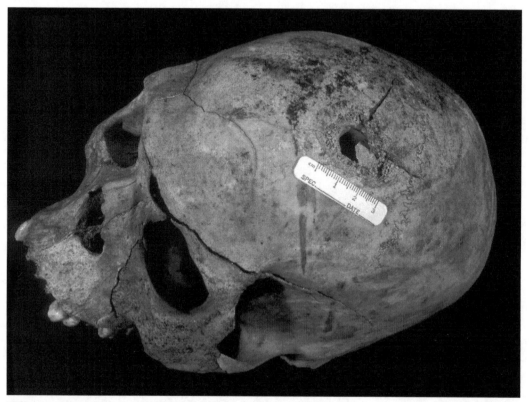

■FIGURE 7: *A penetrating gunshot wound. The projectile can be seen lodged against the cranial vault. Its energy produced external beveling. Photograph by Michael Warren, courtesy of Pictures of Record, Inc., 2001.*

Trajectory

Trajectory is the path the projectile takes through the body and the general relationship between the victim and the shooter. The trajectory of a projectile is almost always non-ambiguous, with the projectile marking the bone on the

Bare Bones: A Survey of Forensic Anthropology

exit side with a cone of energy that marks its direction **(Figure 8)**. This cone of energy produces a phenomenon called "beveling." Entry wounds are normally smaller, with well-defined, punched-out edges. Exit wounds are usually larger, less defined, and characterized by chips in the bone. In the cranium, the entry will be internally beveled and the exit will be externally beveled.

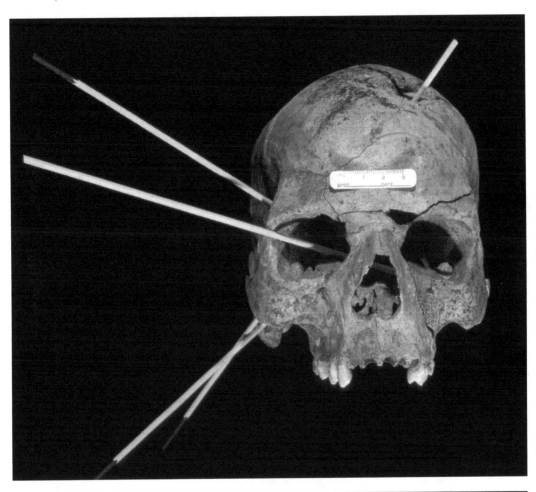

■**FIGURE 8**: *A cranium with multiple gunshot wounds, from multiple trajectories. Anthropologists often use wooden dowels to demonstrate the suspected path of each projectile. Photograph by Michael Warren, courtesy of Pictures of Record, Inc., 2001.*

While the trajectory provides information on the relationship of the victim to the shooter, this is not to say that the anthropologist is always able to say in what position the victim was when he was shot (*e.g.*, was the victim running away, kneeling or attacking the shooter?). Trajectory only tells us which path the projectile took into and through the body **(Figure 9)**.

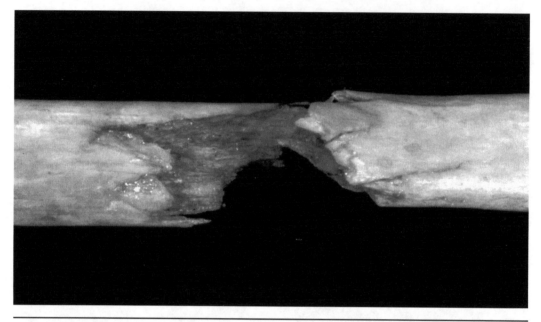

■FIGURE 9: *A gunshot defect that demonstrates internal beveling of a rib.*

Caliber

The caliber of a bullet is the cross-sectional diameter, usually stated in millimeters or a fraction of an inch. So, a 9 mm. projectile is approximately 9 mm. in diameter and a .22 caliber projectile is .22 inches (or 5.56 mm.) in diameter. If a bullet retains its initial diameter as it strikes and passes through tissues, then the entry wound would be expected to measure the width of the round. However, several factors affect this assumption, including range, projectile type and design, and the presence of intermediate targets. So, while it is difficult for the examiner to determine the precise caliber from the diameter of the entry wound, the entry defect often allows one to place the projectile into groups of calibers, such that a defect measuring approximately 9 millimeters in diameter can be expected to have been produced by one of several weapons (*e.g.*, 9 mm., .38, .357 or .380 calibers) provided that the projectile did not expand significantly in the soft tissue prior to impacting the bone (Ross, 1996; Berryman *et al.*, 1995).

Bullet type

A projectile's shape and construction are engineered to achieve a specific goal. For example, target rounds used at a shooting range often have a blunt nose with sharp margins. These "wad cutters" are intended to produce a clear, round hole in a paper target. Since these rounds are used at relatively close range, their non-aerodynamic shape is not an issue. Bullets designed for hunting are made to expand on impact, causing more organ and tissue damage and allowing the projectile to expend most of its kinetic energy into the target. Military rounds and bullets intended for semi-automatic and automatic weapons are *full-jacketed* bullets, meaning that the core of the bullet is surrounded by a copper or steel outer layer. This layer allows the bullet to retain its shape prior to being introduced into the firing chamber (which prevents jamming) and, in the case of steel - or heavier metals such as depleted uranium - jackets, gives some military rounds light armor-piercing capabilities. These jacketed bullets tend to retain their shape after having passed into or through human targets (**Figures 10a and 10b**).

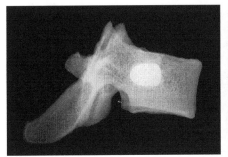 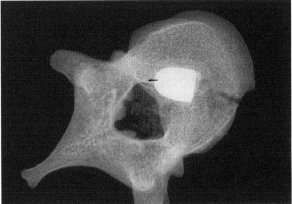

■**FIGURE 10A AND 10B:** *Lateral and superior radiographs of a thoracic vertebra with an embedded 9 mm. projectile. Note that the bullet has retained its shape after passing through bone.*

Velocity

The velocity and energy of a projectile is only discernable within broad categories, such as low velocity/low energy vs. high velocity/high energy rounds. Projectiles striking a target create a permanent track – that is, a cylindrical path that approximates the diameter of the round. However, projectiles that travel over 2200 feet/second or so, create a temporary cavity, via a shock wave, that may produce damage to adjacent tissues, including bone, that do not lie directly along the permanent track of the bullet. A *.22 short* bullet and the *NATO 5.56 mm.* round are essentially the same caliber. But a *.22 short* is a slower-moving, less-energetic round that rarely produces significant damage to surrounding bone. The 5.56 mm., on the other hand, is an extremely fast-moving round with a surprising amount of energy that often creates extensive fragmentation of bone, particularly in the cranial vault.

Statements about velocity must be conservative. The construction of the round, its flight characteristics and other factors render the above statements overly simplistic. But in some cases, the nature of the defect and associated damage can have probative value in helping the anthropologist and pathologist interpret the possible type of weapon used.

Intermediate targets

One confounding circumstance that may alter the appearance of a projectile defect, and therefore alter the interpretation of the trauma, is the possibility of the bullet having passed through a target prior to striking the victim. This intermediate target can be a wall or window, or another victim. Once the projectile has struck an intermediate target, its flight characteristics may change and its diameter may be altered (*e.g.*, it may have enlarged or fragmented into multiple projectiles). The possibility of an intermediate target must be considered when the gunshot-wound defect is atypical, or if the projectile has "behaved" in an uncharacteristic way.

In cases of mass homicide like those seen during ethnic cleansing, other data must be considered in determining the cause and manner of death. The primary issue is whether the decedents in the mass grave are fallen military combatants or murdered civilians (Warren, 2007). The patterns of gunshot injuries are useful in this regard. Among the factors to be considered are:

- The number of wounds per person.

- The number of projectiles embedded within each body (i.e., penetrating wounds).

- The location of the gunshot wounds.

- The existence of multiple trajectories.

- The ratio of wounded to dead.

Certain types of gunshot wounds require special mention. One such lesion, called a *keyhole lesion*, is produced when bone is struck by a projectile at an acute angle. The tangential trajectory produces an entry with well-defined edges and an exit with externally beveled margins within the same defect. The result is a bony lesion that resembles a keyhole. **(Figure 11)**. A keyhole lesion provides firm evidence for the trajectory of the projectile through the cranium (Dixon, 1982). A famous example in which a keyhole lesion provided evidence for trajectory is the case of the assassination of President John F. Kennedy, who was killed as he rode in a motorcade procession through the streets of Dallas, Texas in 1963. The gunshot that struck his head created a massive keyhole lesion that conclusively confirmed that the shot came from behind and above – the fourth floor of the Texas Schoolbook Depository and not the "Grassy Knoll" to the front and slightly above street level.

■**FIGURE 11:** *A skull,* in situ, *with two gunshot defects -- a keyhole lesion and round entry wound to the frontal bone.*

References

Adams JC, Hamblen DL (1992) *Outline of Fractures Including Joint Injuries*, 11th ed. Edinburgh: Churchill Livingstone.

Bartelink EJ, Wiersema JM, Demaree RS (2001) Quantitative analysis of sharp-force trauma: an application of scanning electron microscopy in forensic anthropology. *Journal of Forensic Sciences* 46(6): 1288-1293 .

Berryman HE, Smith OC, Symes SA (1995) Diameter of cranial gunshot wounds as a function of bullet caliber. *Journal of Forensic Sciences* 40(5):751-754.

Clark EGI, Sperry KL (1992) Distinctive blunt force injuries caused by a crescent wrench. *Journal of Forensic Sciences* 37(4):1172-1178.

DiMaio VJM (1998) Gunshot Wounds: Practical Aspects of Firearms, Ballistics, and Forensic Techniques, 2nd Edition. Boca Raton, Fla.: CRC Press.

Dixon DS (1982) Keyhole lesions in gunshot wounds of the skull and direction of fire. *Journal of Forensic Sciences* 27:555-566.

Freas LE (2006) Scanning electron microscopy of saw marks in bone: Assessment of wear-related features of the kerf wall. *Proceedings of the American Academy of Forensic Sciences* 12:296.

Krogman MW (1962) *The Human Skeleton in Forensic Medicine*. Springfield, Ill.: Charles C. Thomas.

Maples WR (1986) Trauma analysis by the forensic anthropologist. In Reichs KJ (ed.) *Forensic Osteology: Advances in the Identification of Human Remains*. Springfield, Ill.: Charles C. Thomas, pp. 218-228.

Ross AH (1996) Caliber estimation from cranial entrance defect measurements. *Journal of Forensic Sciences* 41(4):629-633.

Stewart TD (1979) *Essentials of Forensic Anthropology: Especially as Developed in the United States*. Springfield, Ill.: Charles C. Thomas.

Symes SA, Berryman HE, Smith OC (1998) Saw Marks in Bone: Introduction and Examination of Residual Kerf Contour. In Reichs KJ (ed.) *Forensic Osteology: Advances in the Identification of Human Remains*, 2nd ed. Springfield, Ill.: Charles C. Thomas.

Symes SA (1992) Morphology of saw marks in human bone: Identification of class characteristics. *Dissertation*. Knoxville (TN): University of Tennessee, 1992.

Ubelaker DH (1992) Hyoid fracture and strangulation. *Journal of Forensic Sciences* 37(5):1216-1222.

Warren MW (2007) Interpreting gunshot wounds in the Balkans: Evidence for genocide. Brickley M and Ferllini R (eds.) *Forensic Anthropology: Case Studies From Europe*. Charles C. Thomas. Chapter 10:151-164.

Warren MW (2001) *Introduction to Forensic Anthropology: Human identification and trauma analysis*. Weston, CT: Pictures of Record.

Sample Test Questions

1. **Incised and puncture wounds are examples of:**

   a. Sharp force trauma

   b. Gunshot wounds

   c. Blunt force trauma

   d. Saturday night in a raucous small town

2. **Most civilian bullets designed for "home protection", are designed to:**

   a. Expand on impact

   b. Expend all of their kinetic energy in the target

   c. Penetrate 12 to 18 inches into ballistic gelatin

   d. All of the above

3. **Bones tend to break in predictable ways, producing "common" types of named fractures. Why?**

   a. Since all people perform similar tasks, certain mechanisms of injury are relatively common

   b. All humans are biomechanically vulnerable in the same ways

   c. A and B

   d. None of the above

4. **A patterned injury can help determine:**

   a. The class of weapon or tool used

   b. If the victim was conscious during the assault

   c. The exact weapon or tool used

   d. Whether the assailant was trying to kill or simply injure the victim

5. **Internal and external beveling produced in the cranial vault permits the investigator to determine:**

a. The trajectory of the bullet

b. The caliber of the bullet

c. The distance, or range, that the victim was from the shooter

d. B and C

Burned and Cremated Bodies

A burned body is found among the ruins of a house fire. Are the remains from someone who was killed by smoke inhalation or flames; or is it the body of a murder victim whose assailant is attempting to cover up evidence of his or her crime? What can be learned from the examination of remains that have been burned? When bodies have been burned beyond recognition, what evidence is left that permits scientists to identify the remains? Forensic anthropologists are often asked to weigh in on cases in which human remains have been burned, to assist the pathologist with identification of the victim and also to help differentiate between perimortem trauma and postmortem thermal alteration of the body.

In almost every case, bodies exposed to structure fires and to car fires retain much of their underlying tissues (Bass, 1984). As a body burns, the muscles contract. Since the flexor muscles of the body are stronger than the extensor muscles, the extremities curl up into what has been described as a pugilistic, or fighting posture. The skeleton is protected by deep layers of muscle and connective tissues in most places, but becomes vulnerable at the joints of the elbows, wrists and knees. However, even the most intense house and car fires rarely damage the skeleton to the extent that the bones become completely calcinated (*e.g.*, lose their organic components). Much of the skeletal evidence used by forensic anthropologists is still intact in burned bodies, and the same techniques used for identification are still available (*e.g.*, antemortem-postmortem radiographic comparison, DNA, and morphological indicators).

A crucial first step in the investigation of a fire-related death is the recovery of the remains. Charred and calcined bone is difficult to spy among similarly colored fire debris unless the investigator has expertise in osteology. Crime scene investigators, fire fighters and law enforcement personnel may not recognize irregular fragments, such as those of the cranial vault, which are critical pieces of the puzzle **(Figure 1)**. These pieces must be carefully reconstructed in the

laboratory so the anthropologist can recognize patterned injuries and differentiate perimortem trauma from postmortem damage secondary to the fire, structural collapse, efforts to extinguish the flames, and other confounding factors (Pope, 2005; Symes *et al.*, 2005). Reconstruction of the skeleton, particularly the skull, is vital. The reconstructed skull may clearly demonstrate patterned trauma which is not evident when the fragments are viewed separately **(Figure 2)**.

Bare Bones: A Survey of Forensic Anthropology

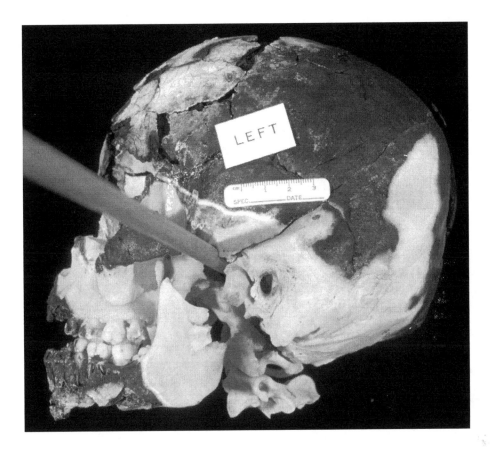

Perimortem trauma vs. postmortem damage

It is not uncommon for skeletal remains to be found following a woods or brush fire, which burns underlying vegetation away and exposes the body so that it can be seen. Investigators must then determine whether the skeleton was already present when the fire swept over the area, or whether the skeleton represents the remains of someone who was killed in the fire, leaving only the skeleton to be discovered.

Bone responds differently to fire depending on whether it is in the fleshed ("green") or dry state. Bone expands at the microscopic level as it heats up; then, having lost its moisture, shrinks as it cools (Rhine, 1998; Nelson, 1992). Some researchers have suggested that bones may shrink up to 25%, which can affect sex and age determinations that rely on size instead of shape (Ubelaker, 1999; Heglar, 1984).

Burned and Cremated Bodies

Dry bones are devoid of or deficient in organic collagens and are unprotected by overlying tissues. Therefore, they expand and contract evenly, usually creating longitudinal cracks and/or uniform "checking" (Nelson, 1992). Bones burned in the flesh are exposed to uneven heating and are prone to extreme warping and transverse, cup-shaped fractures as the fire advances up the bone. The direction that the fire progressed along the body can be discerned by the directionality of these cupped fractures **(Figure 3)**. By knowing in advance the areas of the body that are most vulnerable to fire, experts can determine whether the body was constrained during combustion, or free to contract in the normal way (Symes *et al.*, 2005).

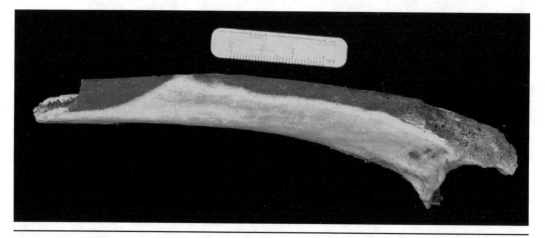

FIGURE 3: *This ulna has only slight burn damage. The margins of the burned areas show where the bone was protected by overlying connective tissues.*

The color of burned bones also tells a story. Bones progress from their normal yellow-brown color to charred black, then gray and white, as they become calcinated (Heglar, 1984). Various classification schemes have been used to categorize burned bone and to discover the relationship of their color to the temperature and duration of the fire (*e.g.*, Glassman and Crow, 1995; Baby, 1954; Table 1).

In summary, burned bones pose a tremendous challenge to forensic scientists. We are continuing to learn more about the behavior of bones exposed to fire. Anthropologists can play a crucial role in these types of cases!

Bare Bones: A Survey of Forensic Anthropology

TABLE 1: *Crow-Glassman Scale for the classification of burned bones (after Correia and Beattie, 2002; Glassman and Crow, 1995)*

Type	Subtype
1[a]	Unburned; no thermal trauma
1[b]	Charred; only minor thermal trauma
2	Partially cremated; internal organs present and some of the limbs
3	Internal organs present and calcined limb fragments
4	Internal organs present, but mainly calcined skeletal tissue
5	Incompletely cremated; calcined skeletal tissue
6	Completely cremated; only ashes remaining

CONTEMPORARY COMMERCIAL CREMATION

A more intense, deliberate type of burning is seen in contemporary commercial cremation. Cremation ranks second only to burial as the preferred method of disposal of human remains in the United States. People choose cremation for a variety of reasons: it is generally less expensive than burial; it is more environmentally-friendly than placing embalmed bodies into the ground; and cremains – otherwise known as cremated human remains - can be inurned, placed in a niche, buried, or scattered. How are cremains scattered? Families can choose to have their loved one's cremains carried aloft in a helium balloon; scattered from an aircraft over an ocean or volcano; and in the United Kingdom, a company called *Heavens Above Fireworks*, even launches cremains into the sky as a fireworks display (Robbins, 2007)! In 1997, the cremains of Gene Roddenberry, creator of the "Star Trek" series, became the first to be shot into space, where they remain in orbit aboard a satellite. A small amount of the cremated remains of astronaut Gordon Cooper, actor James Doohan (who portrayed "Scotty" on the Star Trek television series) and about 200 others were launched into suborbital space aboard a Spaceloft XL rocket in 2007 (Associated Press, 2007).

Over the past two decades, commercially cremated remains have become the subject of litigation in which the plaintiffs have charged that the remains in their possession were either not those of their loved one, or contained the remains of more than one individual. The primary reasons behind the disputes are related to the fact that commingling is inevitable in every cremation.

Burned and Cremated Bodies

Cremation retorts, or furnaces, retorts are designed in a way that makes it difficult to effectively remove all of the remains following a cremation. From a theoretical perspective, **Locard's Principle** (or, the *Theory of Transfer*) tells us that we should expect some degree of commingling in every case. However, the amount of acceptable commingling has not been fully established by either the industry or the courts. Also, cremation – at least at the high frequency seen today – is still relatively new to the United States. Cremation procedures are still largely unregulated, with crematory operators generally "certified" by the manufacturer of the equipment.

Cremated human remains have been studied since the 1960s, when anthropologists realized that, despite the reduction and destruction caused during the cremation process, much could be learned by close examination and research. These early studies were mostly concerned with interpreting prehistoric cremation practices and rituals (Merbs, 1967; Binford, 1963; Wells, 1960; Baby, 1954). Several recent books have discussed the expanding role of forensic anthropologists in cremains investigation, many citing highly publicized incidents involving crematoria using questionable standards (Bass and Jefferson, 2003; Iverson, 2001; Willey and Scott, 1999; Maples and Browning, 1994). Case studies have also been presented in scientific journals and professional meetings (Warren *et al.*, 1999; Murad, 1998; Kennedy, 1996; Murray and Rose, 1993).

Attorneys handling lawsuits dealing with cases of disputed identity, negligent cremation practices and excessive commingling of more than one decedent have sought expert help in order to aid the court's decisions. Several class-action suits of note have involved literally hundreds of plaintiffs, each with potentially millions of dollars at stake. The cremation industry, and mortuary industry in general, have been placed under public scrutiny as industry standards and practices are developed and instituted. Specific legal issues, such as establishing an "acceptable" amount of commingling, are presently being determined by the courts and we can expect this process to continue over the next several years.

Because cremains are principally composed of calcined fragments of bones and teeth, the expert usually takes the form of a forensic anthropologist, a specialist in skeletal anatomy, morphology, and taphonomy (literally, *burial laws*, but in a biological sense the processes that occur to a body after death). Ideally, both an anthropologist and a forensic odontologist should work

Bare Bones: A Survey of Forensic Anthropology

together in these cases. An odontologist should certainly be consulted if dental evidence is found among the cremains. In some major cases, a multidisciplinary team has been retained, including an anthropologist, a pathologist, and an odontologist – each bringing a different area of expertise to bear.

The goals of cremains analysis are to: (a) confirm that the cremains are those of a human; (b) confirm that the remains primarily represent one body and are not significantly commingled; and (c) to establish that the remains are most likely those of a specific individual. Identification in cremation cases is based on the preponderance of presumptive evidence available, as described in Chapter 9. In almost every cremation case, no positive lines of evidence for identity exist. In other words, an examination may reveal that the cremains are *most likely* those of a given individual. In a case where multiple lines of evidence are found, the examiner will provide an expert opinion that the evidence either supports or does not support an identification.

The anthropologist performs an examination of the cremated remains using microscopy, radiography, and other methods to find clues that might lead to identification. These clues generally come from three sources: (1) examination of the biological remains, consisting of osseous (bony) and dental fragments; (2) discovery and identification of non-osseous artifacts; and (3) the elemental/chemical profile of the remains.

THE BIOLOGICAL REMAINS

A proper cremation cycle results in the complete incineration of the organic parts of the body. Of forensic importance is the loss of DNA, one of the most valuable positive lines of evidence for personal identity in a forensic context. While it is possible to extract DNA from badly burned remains (such as those from a house fire or plane crash), commercial cremation completely destroys all of the organic constituents of the body, leaving only the inorganic components of the skeleton and dentition for examination. This material is further processed until the biological component of cremated remains consists of only of thousands of small, calcined bone and tooth fragments. The fragments are usually so small that even an accomplished human osteologist has difficulty in recognizing only a few small fragments in each urn, apart from simply stating that the remains are fragments of bone and teeth.

With the exception of an occasional intact ear ossicle (the very small bones of the middle ear) almost every fragment is non-diagnostic in terms of which bone is represented, and whether or not the bone is human in origin. Therefore, only general statements are offered about the biological remains - either they are consistent with known information about the decedent, or they are not.

Obviously, the more completely remains are processed, the less diagnostic the osseous fragments will be. Much depends on the type of processor (variously called a pulverizer or cremulator) used. Older types of processors, including hammer-mill processors may leave a host of diagnostic osseous fragments. Newer rotary blade processors, provided the blade is not worn and still retains the "lifting" portion of the distal ends, are the most thorough of the processor types. However, even the new processors do a poor job if the blades are worn (Warren and Schultz, 2002).

Cremation Weights

The weight of cremains can provide some presumptive data about the decedent, provided that all of the cremains are present at the time of examination. The expected range of the weight of cremated remains is well-documented and loosely correlated with such variables as cadaver stature and estimated skeletal weight (Warren and Maples, 1997; Tables 2-4). If the weight of the cremains exceeds the ranges established by researchers, then the examiner must consider the possibility that more than one individual is represented (*i.e.*, the cremains are commingled). Weights less than the published ranges may indicate remains that are either incomplete (*i.e.* some portion of the remains have been removed prior to examination), or the remains represent those of a long bone donor, a juvenile or a non-human (pets can also be cremated). If all of the cremains are present, the weight should correspond with published values for the sex and general skeletal robusticity of the decedent (Van Deest, 2007, Bass and Jantz, 2004; Warren and Maples, 1997; Sonek, 1992; Birkby, 1991).

Male:	$y = 42.08$	(V^i)	—	4497.00	±	383.08
Female:	$y = 39.09$	(V^i)	—	4410.00	±	293.95
Pooled:	$y = 54.83$	(V^i)	—	6820.00	±	371.10

Bare Bones: A Survey of Forensic Anthropology

TABLE 2: *Regression formulae for predicting cremains weight (y) from cadaveric weight (V^1). Warren and Maples (1997).*

Male:	$y =$ 42.08	(V^2)	–	4497.00	±	383.08
Female:	$y =$ 39.09	(V^2)	–	4410.00	±	293.95
Pooled:	$y =$ 54.83	(V^2)	–	6820.00	±	371.10

TABLE 3: *Regression formulae for predicting cremains weight (y) from cadaveric stature (V^2). Warren and Maples (1997).*

Male:	$y =$ .1807	(V^3)	+	1136.90	±	435.50
Female:	$y =$ .1920	(V^3)	+	556.34	±	330.99
Pooled:	$y =$ .2791	(V^3)	+	89.284	±	420.52

More research is needed to examine the full range of expected cremains weights from a number of different crematories, since it has been demonstrated that differences in mean weights exist from crematory to crematory (Table 5)(Van Deest, 2007a, 2007b; Bass and Jantz, 2004; Warren and Maples, 1997). The differences are probably the result of differing equipment and procedures; however, some researchers have explored the possibility that regional differences in body composition related to diet may affect cremains weight (Bass and Jantz, 2004).

TABLE 4: *Regression formulae for predicting cremains weight (y) from calculated skeletal weight (V3). Warren and Maples (1997).*

		MALES			FEMALES	
STUDY	N	MEAN	S.D.	N	MEAN	S.D.
Sonek	76	2801.38	589.47	63	1874.87	528.82
Warren & Maples	50	2898.70	499.20	40	1829.38	406.53
Bass & Jantz	151	3379.77	634.98	155	2350.17	536.43
Van Deest	90	3215.56	687.89	99	2247.79	506.10

Osseous Material

Bone and dental fragments are located and identified with the aid of light microscopy and radiography. Most investigators use standard testing sieves to segregate the fragments and particles into uniform sizes (*e.g.*, 4 mm., 2 mm., 1 mm. and < 1 mm.), which allows one to search for diagnostic materials and determine the type of processor used during the reduction of the cremains. Bone and tooth fragments might provide clues that the cremains are those of a human; establish whether or not the decedent had dentition or prosthetic dental work; and, in some cases, detect the presence of a specific age-related pathology (*e.g.*, Mönckeberg's sclerosis; Warren *et al*. 1999).

CREMATION ARTIFACTS

The non-osseous artifacts that are invariably found among cremains can be the most important diagnostic material used to establish identity. For purposes of classification, cremation artifacts have been divided into five categories by Warren and Schultz (2002):

- Medical

- Dental

- Mortuary

- Personal

- Miscellaneous

Medical artifacts might include surgical staples, wire sutures, vascular clips, embolism filters, and pacemaker wires. Dental artifacts generally consist of metallic crowns, posts, bridgework and porcelain crowns and caps. The most common mortuary artifact is called an injector needle – a small "nail" that is used to wire the maxilla and mandible together. Personal artifacts include any materials that were worn by the decedent or introduced into the cremation container prior to cremation. The miscellaneous category is a catch-all and would include cremation slag, which can best be described as amorphous, unidentifiable materials and metallic debris.

CHEMICAL TESTING

The contents of an urn can also be chemically tested to confirm whether they consist of cremated bone and not foreign material. Laboratory tests that have been used for elemental analysis of cremated remains include proton-induced X-ray emission (PIXE), inductively coupled plasma mass spectrometry (ICP-MS), and inductively coupled plasma optical emission spectroscopy (ICP-OES or ICP-AES).

The major elements normally found in bone should be present in cremains, including calcium (Ca), phosphorous (P), potassium (K), sodium (Na), magnesium (Mg), iron (Fe), and a few other essential elements. Elements not found in large volume in bone or other human tissues should be absent. In one case, a family member of the decedent substituted sand for human cremains, which produced high levels of silicates inconsistent with levels found in biological tissues (Warren *et al.*, 2002). Schultz and colleagues (2005) found examples of a rare earth element, gadolinium (Gd) in the cremains of decedents who had undergone medical diagnostic imaging. Gadolinium is a common rare earth element introduced into the body as a contrast medium, enhancing targeted tissues during radiologic imaging procedures (Kirchin 2003, Gibby *et al.*, 2004). Other investigators have found elevated levels of lead in decedents who had been shot and carried the projectile in their bodies for an extended period of time (Bodkin *et al.*, 2005). More research needs to be conducted, but these preliminary investigations show that elemental/chemical analyses can inform anthropologists about certain aspects of the decedent's life history.

The use of chemical methods for elemental analysis, including ICP-MS, XRF, PIXE, ICP-OES, ICP-AES and other methods should now be considered as an integral part of a thorough cremains investigation. Chemical analyses can answer elementary questions about the composition of cremains that cannot be answered from gross methods alone. Chemical methods can be used to determine whether the disputed cremains are comprised of bone, a foreign material substituted for cremains, or both.

Personal identification

When establishing identity from cremains, as in any case involving death investigations of decedents of unknown identity, antemortem data must be gathered about the putative decedent. As discussed in previous chapters, the collection of antemortem data from the family members of the decedent is a familiar process to anthropologists working in human identification. Among the data that might be helpful to the investigator are:

- The decedent's biological sex

- The decedent's age at death

- The decedent's cause and manner of death

- The decedent's medical and dental history

- The decedent's mortuary history (*e.g.*, was the body embalmed prior to cremation? Was the body clothed? Were any foreign items placed in the cremation container?).

Armed with this antemortem information, the anthropologist can begin to compile a life history for the putative decedent. The investigator can then determine whether her findings correspond with the provided antemortem data. For example, if the family reports that their loved one had no remaining natural dentition, but instead wore dentures, then a finding of several tooth roots and crowns would be problematic in the identification of the decedent's cremains.

References

Associated Press (2007) Rocket carries ashes of 'Star Trek' star. *Gainesville Sun*, April 29, 2007.

Baby RS (1954) Hopewell cremation practices. *Ohio Historical Society Papers in Archaeology* 1:1-17.

Bass WM (1984) Is it possible to consume a body completely in a fire? In Rathbun TA and Buikstra JE (eds.) *Human Identification: Case Studies in Forensic Anthropology*. Springfield, Ill.: Charles C. Thomas. p 159-167.

Bass WM, Jantz RL (2004) Cremation weights in east Tennessee. *Journal of Forensic Science* 49(5):901-904.

Bass WM and Jefferson J (2003) *Death's acre: inside the legendary forensic lab, the Body Farm, where the dead do tell tales*. New York: G.P. Putnam's Sons.

Binford LR (1963) Analysis of cremations from three Michigan sites. *Wisconsin Archeologist* 44:98-110.

Birkby WH (1991) The Analysis of Cremains. Paper Presented at the 43rd Annual Meeting of the *American Academy of Forensic Sciences*, Anaheim, CA, February 18-23, 1991.

Bodkin T, Potts G, Brooks T, Shurtz K (2005) Elemental analysis of human cremains using inductively coupled plasma optical emissions spectroscopy (ICP-OES) to distinguish between legitimate and contaminated cremains. *Proceedings of the American Academy of Forensic Sciences* 11:307.

Correia PMM, Beattie O (1997) A critical look at methods for recovering, evaluating, and interpreting cremated human remains. In Haglund WD, Sorg MH (eds.) *Advances in Forensic Taphonomy: Method, Theory, and Archaeological Perspectives*. Boca Raton: CRC Press, pp. 435-450.

Gibby WA, Gibby KA, Gibby WA (2003) Comparison of Gd DTPA-BMA (Omniscan) versus Gd HP_DO3A (ProHance) retention in human bone tissue by inductively coupled plasma atomic emission spectroscopy. *Investigative Radiology* 39:138-142.

Glassman DM, Crow RM (1995) Standardization model for describing the extent of burn injury to human remains. *Journal of Forensic Sciences* 41(1):152-154.

Heglar R (1984) Burned Remains. In Rathbun TA and Buikstra JE (eds.) *Human*

Identification: Case Studies in Forensic Anthropology. Springfield, Ill.: Charles C. Thomas. p 148-158.

Iverson KV (2001) *Death to dust: what happens to dead bodies?*, second edition. Tucson, AZ: Galen Press, Ltd.

Kennedy KA (1996) The wrong urn: commingling of cremains in mortuary practices. *Journal of Forensic Science* 41:689-692.

Kirchin MA (2003) Gadolinium contrast agents for MR imaging: Safety update. *Highlights in MRI* 1:1-3.

Maples WR, Browning M (1994) *Dead men do tell tales: the strange and fascinating cases of a forensic anthropologist.* New York: Broadway Books.

Merbs C (1967) Cremated human remains from Point if Pines, Arizona. *American Antiquity* 32:498-506.

Murad T (1998) The growing popularity of cremation versus inhumation: some forensic implications. In Reichs KJ (ed.) Forensic Osteology: Advances in the Identification of Human Remains. Springfield, Ill.: Charles C. Thomas. p 86-105.

Murray KA, Rose JC (1993) The analysis of cremains: a case study involving the inappropriate disposal of mortuary remains. *Journal of Forensic Science* 38:98-103.

Nelson R (1992) A microscopic comparison of fresh and burned bone. *Journal of Forensic Sciences* 37(4):1055-1060.

Pope EJ (2005) Utilizing taphonomy and context to distinguish perimortem from postmortem trauma in fire deaths. *Proceedings of the American Academy of Forensic Sciences* 11:331.

Rhine S (1998) *Bone Voyage: A Journey in Forensic Anthropology.* Albuquerque: University of New Mexico Press, p. 7.

Robbins J (2007) Roadblock for Spreading of Human Ashes in Wilderness. *The New York Times*, national edition, A19.

Schultz JJ, Warren MW, Krigbaum JS. 2005. Analysis of modern cremated human remains: gross and chemical methods. *American Journal of Physical Anthropology*, Supplement 40:184.

Sonek A (1992) The weight(s) of cremated remains. Paper presented at the 44th annual meeting of the *American Academy of Forensic Sciences*, New Orleans, LA.

Symes SA, Woytash JJ, Kroman AM, Wilson AC (2005) Perimortem bone fracture distinguished from postmortem fire trauma: A case study with mixed signals. *Proceedings of the American Academy of Forensic Sciences* 11:288-289.

Bare Bones: A Survey of Forensic Anthropology

Ubelaker DH (1999) Human Skeletal Remains: Excavation, Analysis, Interpretation. Washington, D.C.: Taraxacum Press.

Van Deest TL (2007a) Sifting through the "ashes": Age and sex estimation based on cremains weight. *Proceedings of the American Academy of Forensic Sciences* 13:378-379.

Van Deest TL (2007b) Sifting through the "ashes": Age and sex estimation based on cremains weight. *Masters thesis, California State University at Chico.*

Warren MW, Falsetti AB, Hamilton WF, Levine LJ (1999) Evidence of arteriosclerosis in cremated remains. *American Journal of Forensic Medicine and Pathology* 20:277-280.

Warren MW, Falsetti AB, Kravchenko II, Dunnam FE, Van Rinsvelt HA, Maples WR (2002) Elemental analysis of bone: proton-induced X-ray emission testing in forensic cases. *Forensic Science International* 125:37-41.

Warren MW, Schultz JJ. 2002. Post-cremation taphonomy and artifact preservation. Journal of Forensic Science 47:656-659.

Warren MW, Maples WR (1997) The anthropometry of contemporary commercial cremation. *Journal of Forensic Science* 42:417-423.

Wells C (1960) The study of cremation. *Antiquity* 34:29-37.

Willey P, Scott DD (1999) Clinkers on the Little Bighorn Battlefield: in situ investigation of scattered recent remains. In Fairgrieve SI (ed.) *Forensic Osteological Analysis: A Book of Case Studies.* Springfield, Ill.: Charles C. Thomas.

Sample Test Questions

1. **Why does the law specify that a 48-hour period must pass before a body can be cremated?**

    a. To allow family members time to view the body before cremation

    b. So OSHA can inspect the cremation facilities prior to cremation

    c. To prevent the destruction of biological evidence before it can be examined by experts

    d. Most religions prohibit direct, immediate cremation of human remains

2. **What are the reasons for disputed remains?**

    a. Some commingling is inevitable and may be discovered by the family of the decedent

    b. Some crematoria might use poor practice and procedures

    c. There is a lack of standards in the cremation industry

    d. All of the above

3. **Several vascular clamps, or ligation clips are found among cremains. The decedent has no surgical history.**

    a. The cremains are those of another decedent

    b. The ligation clips are one line of presumptive evidence that commingling occurred during cremation

    c. The ligation clips are proof that the cremationist did not follow standard procedures

    d. The ligation clips have no probative value

Mass Fatalities

The most important effort we can make immediately following a disaster is to locate and treat the sick and injured. In the aftermath of Hurricane Katrina, many people were stranded and in danger, so the dead were given lesser priority while authorities attempted to rescue and evacuate the survivors. However, a mass disaster inevitably causes the deaths of many, and there must be plans in place to recover and identify these victims. Disastrous events that result in large numbers of dead are referred to as **mass disasters** or **mass fatality incidents**. An event might be called a mass disaster when the numbers of injured, sick, dying and dead victims overwhelm local resources. For instance, a bus accident that occurs in a small community may initiate a disaster plan that brings in outside resources and personnel, whereas if the same accident had occurred in a large metropolitan area, adequate manpower and resources may already be available to handle the treatment of the injured; and the removal and identification of the dead.

Mass disasters include famine, epidemics, fires, floods, volcanoes, and other natural events, while the phrase *mass fatality incident* for is often used to describe human-made events such as war, terrorism, riots, aircraft crashes. Whichever definition is used, it is obvious that these events might result in large numbers of dead. Therefore, forensic anthropologists have come to play an increasingly important role in the identification of victims and the investigation of the cause of death of victims of mass fatality incidents due to their expertise in the examination of skeletal, fragmented and burned remains.

GOALS IN MASS DISASTERS

The recovery and examination of bodies following mass disasters is, like cases with single decedents, directed towards personal identification and determination of the cause and manner of death. However, many mass-fatality incidents are different from the forensic anthropologist's normal casework. Most cases examined by forensic anthropologists are individual victims of homicide whose skeletal remains are found at a crime scene. Since the remains can be those of any missing person, the remains are said to be part of an **open population**. Many mass fatality incidents, such as a downed commercial aircraft, represent a **closed population**. As there is a record (the manifest) of all of the passengers presumed to be on board, the universe of possible decedents is relatively small. Identifying characteristics need only be compared to those of known passengers, which gives each line of presumptive evidence greater probative value. A simple example is the crash of a small airplane in which three persons were on board – the male pilot, his wife and their child. Identification of juvenile skeletal remains may have enough presumptive value that a medical examiner would identify those remains as those of the child. In the case of an earthquake, where there are many victims, the finding of juvenile remains – taken alone as presumptive evidence – would have little probative value, as juvenile remains could represent any one of many children who were killed.

While the primary goal in mass disasters is the recovery and identification of the victims, many mass-fatality incidents may also involve an investigation in which the cause and manner of death is also an important factor. The findings of forensic scientists will inevitably affect issues ranging from public safety and aircraft design to the outcome of civil litigation aimed at reparation for victim's families.

Mass disasters often result in burned, fragmented, and scattered remains. Because anthropologists have a long history of research designed to glean information about past populations from scant, partial, human remains, we have become an increasingly important member of the interdisciplinary teams whose job it is to recover, identify and analyze the remains of victims.

PLANNING FOR MASS DISASTERS

Since every disaster is different, how can authorities plan a cohesive response? The public can rest assured that detailed planning will not account for every contingency that *makes disasters disastrous*. But, some broad considerations should be addressed.

First, responders must know if the event involves an open or closed population. This dictates the strategies for instituting the family assistance center and give responders an idea of how much antemortem data they can expect. Did the disaster result in fragmented or complete remains? How will this effect triage? The degree of fragmentation, as well as the number of victims, may determine the role of DNA in the identification process. The Medical Examiner or Coroner will need to make some difficult decisions about identification procedure. Will every individual, or fragment thereof, be identified? What will happen to remains that are not identified? The families should be consulted and must decide relatively early in the identification process how to handle notifications of identification and re-association of multiple body parts.

AT THE SCENE

The first challenge is to devise an appropriate system for search and recovery of remains based on the size and scope of the event (Sledzik and Rodriquez, 2002; Kontanis *et al.*, 2001). Some scenes are contained within a relatively small area (*e.g.*, the Branch Davidian Compound and the Rhode Island nightclub fire). These disasters result in fragmentation and commingling of remains. Although these remains are contained within known boundaries, it is difficult to recognize and differentiate human remains from rubble. It is especially difficult to identify burned and fragmented bones, because they do not look like bones to inexperienced people. Crime-scene personnel often walk right past human remains – particularly burned skeletal remains – because they have mistaken the calcined bone for debris. Other scenes encompass a large area. Both Hurricane Katrina and the Space Shuttle Challenger reentry breakup spanned more than one state. These disasters required extensive searching and mapping strategies in order to increase the efficiency and effectiveness of the recovery effort.

Most agencies, such as DMORT (see below), will send a multidisciplinary team to the scene to provide an initial field assessment of the scale of the disaster, and the condition of the remains, so that the appropriate numbers of experts can be deployed to the scene. Some disasters might require more odontologists and fewer anthropologists; some may require a regional response, while others may demand a national response to the incident (*e.g.*, the September 11th terrorist attacks). Some scenes require special expertise, such as forensic engineers, to assess the integrity of partially collapsed structures. After the bombing of the Alfred P. Murrah building in Oklahoma City, engineers constantly assessed the soundness of the remaining structures to insure that those recovering the remains were safe from harm.

DMORT

The governmental agency tasked with the identification and examination of the bodies of victims of mass disaster is the **Disaster Mortuary Operational Response Team**, or **DMORT**. Initially formed by the National Funeral Directors Association in the early 1980s, DMORT has grown to over 1,200 members. DMORT currently functions as an arm of the National Disaster Medical System, under the auspices of the United States Public Health Service and the Office of Emergency Preparedness (Saul and Saul, 2003; Sledzik and Willcox, 2003). DMORT can be utilized anywhere within the United States and its territories by request of ranking government officials. Deployment of DMORT is part and parcel of federal assistance following disasters and is generally provided without cost to local and state agencies. The role of DMORT is to provide organized plans and personnel who have prior experience in the recovery and identification of victims of mass disaster. DMORT does not supplant the local authority. The Medical Examiner or Coroner still has the legal mandate to identify the victims and issue a death certificate. The experts within DMORT work with the local authority to offer support as needed and requested. This support is tendered in the form of three major resources: DMORT personnel, the **Family Assistance Center** (FAC) and the **Disaster Portable Morgue Unit** (DPMU).

DMORT personnel includes forensic pathologists, forensic anthropologists, forensic odontologists, funeral directors, evidence technicians, DNA technicians, and other experts with prior experience in human identification. Each of these experts is used to working as part of a multidisciplinary team to

Bare Bones: A Survey of Forensic Anthropology

achieve the goal of identification and determination of cause and manner of death. This is a key feature of DMORT, as most local jurisdictions have only a few experts available. In almost every case, these local experts have welcomed the additional help provided by DMORT.

The FAC is perhaps the most important cog in DMORT's operations. Personnel at the FAC are responsible for interacting with family members and collecting antemortem information on the victims that is vital to identification efforts (Brannon and Kessler, 1999; Sledzik and Kontanis, 2005). The FAC also provides a secure location where the family and loved ones of the victims can gather, receive news directly from the proper sources (instead of through the media, which can lead to miscommunication), and receive emotional support from each other and experienced grief counselors.

DMORT currently maintains two DPMUs – one in Maryland and one in California (Sledzik and Kauffman, 2007), and a third morgue is planned. These portable morgues are stores of specialized equipment and supplies ready to be deployed to a disaster site by land, sea and air within 48 hours by a core team of logistical experts known as the *Red Shirts*. The DPMU contains thousands of items, including portable x-ray machines, autopsy equipment, anthropological measuring devices, computers and personal protective equipment (see the DMORT website).

The beauty of the DPMU is its consistency. Anthropologists know that when they arrive at the morgue, their station will have familiar gear and equipment, that it will invariably be situated between the pathologists and the odontologists, and that the radiology station will be within earshot.

AT THE MORGUE

The layout of the DPMU is intuitive and simple. In fact, various forensic experts working around the world have independently adopted similar layouts for use in temporary morgues from Thailand to the Balkans.

The general layout of the morgue channels bodies and body parts through in a specific sequence (see **Figure 1**). First, the remains are brought in from the field and stored in refrigerated containers – in the United States these are often "reefer" trailers attached to semi-tractors. The bodies or body parts are then brought to a triage station, where experts determine whether the remains are likely to yield an identification or not. **Triage** is the process of deciding which remains have the most likelihood of resulting in a personal identification. One role of the anthropologist working in the DPMU is to assist in the triage of the remains. Whole bodies are processed first, followed by larger fragmentary remains. Some fragmentary remains have almost no probative value and are left until all identifiable body parts have been processed. It is likely, in most disasters, that some remains will be left and deemed "common tissue." In triage, the anthropologist separates commingled remains and describes incomplete or fragmented remains for the pathologists.

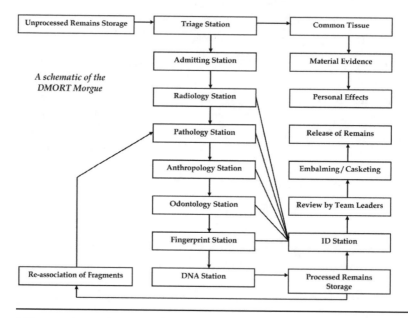

A schematic of the DMORT Morgue

■FIGURE 1: *A schematic of a disaster mortuary.*

Next, the remains are moved to the admitting station where evidence technicians remove material evidence (including personal effects), then assign a tracking number to them. This tracking number is the number of record for the body or body part and is cross-listed with other data, such as information related to provenience. All remains and material evidence are then photographed, described and documented at the admitting station.

The remains then proceed to the radiography station. Intact remains, remains in body bags, and body parts are radiographed by radiology technicians prior

Bare Bones: A Survey of Forensic Anthropology

to further examination. This permits the experts to evaluate the remains for commingling, identifying characteristics, and markers of age and sex. Whole body radiographs also permit the examiners to check for dangerous materials within and surrounding the remains. In war zones, for example, it is best to check for unexploded ordinance before the remains are handled. As bodies and body parts are moved through the morgue, examiners may request a specific radiologic view of a specimen. The DMORT DPMUs currently utilize computed radiography consisting of a portable x-ray machine, digital x-ray plates, and a digital scanner. The radiology technicians bring up the image on a computer screen, which quickly becomes surrounded by pathologists, anthropologists and dentists, each with their own perspective and interests in the view.

The remains then advance to the pathology station. If the remains are relatively intact, the pathologists perform a routine autopsy. As physicians, the pathologists are best qualified to comment on traumatic injuries to soft tissues; and as we will see, it is a pathologist that must determine the cause and manner of death. The pathologists begin to fill out a protocol, complete with various forms, which will follow the remains through the morgue. The anthropologists and odontologists are just one or two stations away, should the pathologist need to consult.

At the anthropology station, workers try to provide a biological profile and to look for unique characteristics that may lead to an identification. The anthropologists work closely with the pathologists and odontologists to compare identifying characteristics with the antemortem data collected at the FAC in order to identify the decedent. Among the duties assigned to the anthropologists are:

- Determine sex, age, race, stature, and distinguishing characteristics.

- Interpret radiographs for age estimation and unique skeletal features.

- Determine the minimum number of individuals based on remains recovered.

- Assist in identifications using anthropological features.

- Analyze trauma evidence and incident related injuries.

- Conduct quality assurance of IDs before release from morgue.

The odontology station follows the anthropology station. Here, experts evaluate the dentition for reparative and cosmetic restorations, dental appliances such as plates and dentures, as well as evaluating palate shape and tooth morphology for clues as to the ancestry of the decedent (**Figure 2**).

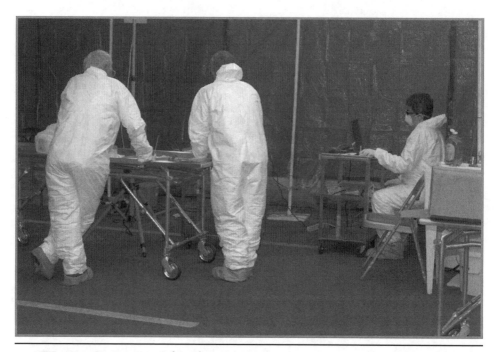

■**FIGURE 2:** *Forensic Odontologists prepare to process remains in the morgue following Hurricane Katrina.*

Odontology is often followed by the fingerprinting station, although in some cases such as commercial aircraft crashes, fingerprinting by the FBI may be done prior to examination by other types of experts. Next, samples of soft tissue, bone and teeth are collected by DNA technicians. These samples are sent to the laboratory where they are compared to latent samples of DNA known to represent the decedent, or if mitochondrial DNA is to be used, maternal relatives who have been contacted through the family assistance center.

Once all of the experts have examined the remains and recorded their findings, the remains are returned to a different refrigerated storage unit. As more and more remains are processed through the morgue, workers may return repeatedly to the storage unit to retrieve associated body parts or review

Bare Bones: A Survey of Forensic Anthropology

findings; especially when more remains are processed that shed light on the identity of remains that have been previously processed.

At any time during the examination of the remains, experts may find evidence that helps to identify the remains. At the end of each day, representatives called *team leaders* from each station meet to review the day's activities and consult on every case in which a tentative identification was made. All team leaders must agree with the identification before the case is presented to the local Medical Examiner. Once the Medical Examiner agrees with the identification, he/she will issue a death certificate and the remains can be returned to the family.

THE WORLD TRADE CENTER ATTACK

On September 11, 2001, two commercial jets were hijacked by terrorists and flown into both towers of the 110-story World Trade Center in Manhattan. The impact, combined with intense heat from burning jet fuel, caused the towers to collapse, resulting in more than 2,700 deaths. The unprecedented forces that occurred during the collapse and burning of the structures produced highly fragmented and altered remains.

New York is one of only a few metropolitan areas in the world that has the resources to plan, organize and support the massive effort required to locate, recover and identify the remains of such a large number of victims. The responsibility of recovering and identifying the victims fell to Dr. Charles S. Hirsch, Office of the Chief Medical Examiner (OCME) for New York City. Bringing his considerable resources to bear, he made an early decision that all remains would be genetically tested; and that, when possible, all identifications would be confirmed by DNA match with latent samples. As noted above, early appraisal of the highly fragmented condition of the remains, consideration of the ability to collect antemortem data, including nuclear DNA exemplars and mtDNA samples from maternal family members, and a large but - in the end - closed population, led Dr. Hirsch to make this crucial, early decision.

This decision has resulted in remarkable results, considering the task at hand. It also altered the roles of many of the team members assigned to the identification efforts – both within the OCME and assisting agencies including DMORT. Although the DPMU had been flown into LaGuardia International

Airport and set up in an abandoned hangar, the facility was never used. Instead, all identification efforts took place at the OCME on First Avenue in Manhattan. Here, anthropologists played an important role, led by Amy Mundorff-Zelson, a forensic anthropologist employed at the OCME, in providing triage skills and identifying commingled and/or fragmented remains. Each identification based on DNA or other criteria was confirmed by the anthropologist by finding corroborating biological evidence that the remains matched the biological profile of the putative decedent (Mundorff-Zelson, 2003).

Anthropologists also contributed to the identification efforts at the Freshkills site. As debris was removed from ground zero it was taken to Staten Island and deposited at the Freshkills Landfill. Human remains invariably were found among the rubble and debris of the towers. Personnel from multiple agencies conducted a systematic search among the debris and were asked to bring any material that might represent human remains to the New York Police Department's Crime Scene Unit, stationed initially in a tent along the edge of the landfill. After several days, the NYPD crime scene technicians and detectives requested that someone with expertise in skeletal anatomy and fragmented remains be deployed to assist in making the determination between human and non-human remains and materials.

Some remains found at the Freshkills site displayed morphological or pathological variation that would have had significant probative value within the context of normal anthropological analysis (*ex.*, myositis ossificans, antemortem fractures). However, the ability to sample every bone for DNA comparison reduced the value of such observations at the site. These characters were no doubt used to confirm the DNA identifications at a later date. What, then, did the anthropologists contribute at Freshkills? We were able to save resources by preventing the DNA technicians from attempting to extract, amplify and sequence DNA from non-human bones and non-biological materials. The World Trade Center was home to several restaurants that produced a surprising amount of non-human bones! The osteological expertise of the anthropologists also reduced unnecessary work and saved valuable time, energy and resources by preventing the evidence technicians from sorting through materials without evidentiary value (Warren, *et al.*, 2003).

At the time of this books writing, identification efforts are still ongoing at the OCME. More human remains were recently found by construction workers clearing and digging for the ground zero memorial and construction of new buildings. Excavations using archaeological techniques are in progress under

Bare Bones: A Survey of Forensic Anthropology

the direction of Drs. Bradley Adams and Christian Crowder, anthropologists for the OCME. With the help of many of their colleagues, they have developed new and efficient ways of sorting through massive amounts of soil and debris in their search for the remains of victims. It is a monumental task.

HURRICANE KATRINA

Hurricane Katrina was a Category 3 hurricane that made landfall near the Mississippi/Louisiana state line on August 29, 2005. The hurricane caused massive destruction, breaching the levees of New Orleans and becoming the costliest natural disaster in United States history. Like the World Trade Center attacks, Hurricane Katrina resulted in large numbers of both missing and dead. However, the differences between the two events are striking and resulted in new challenges for forensic scientists (Fulginiti, *et al.*, 2006). Most of the remains of the Katrina victims were intact, unlike the extensive trauma seen in the WTC victims. However, advanced decomposition due to the extended period of time for many remains to be recovered made the remains of many storm victims unidentifiable using normal means.

The storm took lives over a broad area. Moreover, the survivors and family members of the storm victims were spread over the entire country, confounding attempts to collect and compile antemortem data on the missing. The storm also damaged many of the local medical and dental records of the missing, which otherwise could have been used for antemortem/postmortem comparison.

Hurricane Katrina marked the first time that two portable morgues have been operated simultaneously. One morgue was situated in the Biloxi/Gulfport area of Mississippi and became known as *DMORT East*. The other was located in St. Gabriel, Louisiana, approximately 70 miles west of New Orleans. This morgue, called *DMORT West*, was burdened with the task of identifying the largest number of storm victims. (**Figure 3**).

The DMORT teams and other officials not only had to deal with the recovery and identification of the storm victims, but also were confronted with cemetery disinterments and repatriations. Over 600 caskets became disinterred from both in-ground burials and above-ground crypts. This is not the first time that DMORT has dealt with wayward caskets, but it is the first time that DMORT personnel have processed disaster victims and cemetery remains at the same time!

■FIGURE 3: *DMORT East DMPU. The refrigerated trucks in the distance store processed and unprocessed remains.*

The identification efforts following Katrina lasted over 6 months, with the final number of storm-related dead 1,800 people.

References

Brannon RB and Kessler HP (1999) Problems in mass disaster dental identification: A retrospective review. *Journal of Forensic Sciences* 44(1):123-127.

DMORT website. http://www.dmort.org

Fulginiti LC, Warren MW, Hefner J, Bedore LR, Byrd J, Stefan V, Dirkmaat D (2006) Anthropology Responds to Hurricane Katrina. *Proceedings of the American Academy of Forensic Sciences* 12:305.

Kontanis EJ, Ciaccio FA, Dirkmaat DC, Jumbelic MI (2001) Variables affecting the success of victim identification in mass fatality events, *Proceedings of the American Academy of Forensic Sciences* 7:118.

Mundorff-Zelson A (2003) The role of anthropology during the identification of victims from the World Trade Center disaster. *Proceedings of the American Academy of Forensic Sciences* 9:277.

Saul FP and Saul JM (2003) Planes, trains, and fireworks: The evolving role of the forensic anthropologist in mass fatality incidents, in Steadman DW (ed.) *Hard Evidence: Case Studies in Forensic Anthropology*. Upper Saddle River, New Jersey: Prentice-Hall, chapter 20.

Sledzik PS and Kauffman PJ (2007) The Mass Fatality Incident Morgue: A Laboratory for Disaster Victim Identification. In Warren MW, Walsh-Haney HA, and Freas LE (eds.) *The Forensic Anthropology Laboratory*. Boca Raton, Florida: CRC Press. *In press*.

Sledzik PS and Kontanis EJ (2005) Resolving commingling issues in mass fatality incidents. *Proceedings of the American Academy of Forensic Sciences* 11:311.

Sledzik PS and Rodriguez WC (2002) *Damnum Fatale:* The Taphonomic Fate of Human Remains in Mass Disasters, in Haglund WD and Sorg MH (eds.) *Advances in Forensic Taphonomy. Methods, Theories and Archaeological Perspectives*. Boca Raton, Florida: CRC Press, chapter 17.

Sledzik PS and Willcox AW (2003) *Corpi Aquaticus:* The Hardin Cemetery Flood of 1993, in Steadman DW (ed.) *Hard Evidence: Case Studies in Forensic Anthropology*. Upper Saddle River, New Jersey: Prentice-Hall, chapter 19.

Warren MW, Eisenberg L, Walsh-Haney HA, Saul JM (2003) Anthropology at Fresh Kills: Recovery and identification of the World Trade Center victims. *Proceedings of the American Academy of Forensic Sciences* 9:278.

Sample Test Questions

1. On September 11th, a "DMORT" team was deployed to New York City. What does the acronym "DMORT" stand for?

    a. Disaster Mortuary Operational Response Team

    b. Disaster Medical and Odontological Response Team

    c. Division of Mortuary Operations and Rescue Teams

    d. Division of Military Operations and Recovery Tactics

2. Which of the following experts is not usually a member of a DMORT team?

    a. Forensic Odontologist

    b. Forensic Entomologist

    c. Forensic Pathologist

    d. Forensic Anthropologist

3. Which of the following best describes a mass disaster event?

    a. Plane crash

    b. 5 or more dead or injured

    c. 10 or more dead or injured

    d. More injuries or deaths than local authorities can handle

Human Rights Missions

On December 10, 1948, following the atrocities the world witnessed during World War II, the General Assembly of the United Nations adopted and proclaimed the Universal Declaration of Human Rights. The preamble to that document establishes basic human rights as the "foundation of freedom, justice and peace in the world." (General Assembly resolution 217 A, III). Among those basic rights are the, "right to life, liberty and security of person," as well as the right to know the truth and to reconcile a conflict within the bounds of one's culture. Anthropologists would also add to these the right to one's heritage and history.

Unfortunately, we live in a world where it is all too common for atrocities to take place. Over the last decade, anthropologists, pathologists and other scientists have become increasingly involved in the exhumation and examination of bodies from mass graves filled with victims of genocide and war crimes. The list of countries in which anthropologists and other scientists have investigated abuses of human rights is both astounding and sad: Angola, Argentina, Bosnia, Bolivia, Brazil, Chile, Columbia, Congo, Croatia, East Timor, El Salvador, Ethiopia, Guatemala, Haiti, Indonesia, Iraq, Ivory Coast, Kosovo, Kurdistan, Nicaragua, Panama, Peru, Philippines, Rwanda, Serbia, Sierra Leone, South Africa, Thailand, Uruguay...the list goes on and on.

The beginnings of forensic anthropology's involvement in the scientific investigation of genocide can be traced to Dr. Clyde Snow (see Joyce and Stover, 1991). In 1979, Dr. Snow uncovered and revealed to the world the mass graves of "disappeared" civilians in Argentina. His most important contribution, however, may be his leadership role in training many of the founding members of the Argentine Forensic Anthropology Team (*Equipo Argentino de Antropología Forense*, or EAAF). In 1992, Snow led a group organized by the American Association for the Advancement of Science (AAAS) to Guatemala in order to investigate the mass graves of the *Desapareicids* (the disappeared) – more than 40,000 innocent civilians, including women and children, who had been killed by the Guatemalan military and civil patrols since the 1960s. Snow's experience in Guatemala led to the establishment of the Guatemalan Forensic Anthropology Team. Latin American forensic anthropology teams are among the top practitioners in our field and have garnered the respect of colleagues around the globe. They recently formed an umbrella organization called the Latin American Forensic Anthropology Association (*Asociación Latinoamericana de Antropología Forense, or* ALAF).

Today, there are several non-governmental organizations (NGOs) contributing to the investigation of war crimes in Europe, Africa and other sites around the globe. Among the organizations active in human rights investigations are:

- International Committee for the Red Cross (ICRC)
- Amnesty International
- Physicians for Human Rights (PHR)
- Centre for International Forensic Assistance (CIFA)
- *Medecins Sans Frontieres*
- International Forensic Centre of Excellence (InForce)

The authors strongly encourage students to utilize their education by volunteering for any organization that seeks to make the world a better place in which to live, either locally or globally.

THE GOALS OF THE FORENSIC ANTHROPOLOGIST IN HUMAN-RIGHTS INVESTIGATIONS

The purpose of human-rights investigations are twofold: (1) to collect and record evidence of crimes against humanity; and (2) to identify and return the bodies of the victims to their families and communities. Today, most forensic scientists conducting human rights investigations utilize the **Minnesota**

Protocol, a proposed model for the disinterment and analysis of skeletal remains that includes a comprehensive checklist of the steps in a basic forensic examination (United Nations; ST/CSDHA/12 - 1991 - V. *Model protocol for disinterment and analysis of skeletal remains*). The protocol follows the basic forensic archaeological procedures discussed in Chapter 3, and outlines basic procedures for laboratory analyses that lead to identification and documentation of cause and manner of death. The Minnesota Protocol is useful as a general guide, but inevitably some procedures must be modified to address the ever-present obstacles that arise during every mission.

Exhumation

Evidence begins with context, and in human-rights investigations, context begins with the gravesite. Exhumation of the bodies from mass graves provides investigators with vital information:

- What type of equipment was used to dig the grave?

- What is the time since death of the victims?

- Does the position of the bodies indicate that the victims were killed in a remote location and brought to the gravesite, or were they brought to graveside and then executed?

The excavation of the gravesite is left to forensic archaeologists, who will record and map the site. They take great care to keep each body separate and to limit commingling of remains, which confounds later analyses (**Figure 1**). It is routine for archaeologists to carefully, and with great patience, circumscribe and pedestal the body so that it can be photographed and mapped *in situ*. Finally, the team members record all pertinent information before the body is removed from the grave.

FIGURE 1: *The world watches as members of Physicians for Human Rights excavates a mass gravesite in Bosnia in 1996.*

LABORATORY ANALYSIS

Identify the decedent

After excavation, the remains are brought to a temporary morgue facility for examination (**Figure 2**). All remains are stored in a refrigerated container in order to retard decomposition. Bodies are removed from storage and processed in a logical sequence that will be familiar to readers , because it parallels the morgue flow used during identification of victims of mass disasters as shown in the previous chapter. One exception is the first step, in which the entire body is examined under fluoroscopy. This is a crucial procedure that enables the pathologists and technicians to screen the bodies for unexploded ordinance, planted traps, and other dangers. The fluoroscope also permits the pathologist to visualize projectiles, document their location, and remove them so they can be entered as material evidence. Prior to autopsy, the victim's clothing is removed, described in detail, and photographed. All personal effects are entered into evidence by technicians, who also describe and document each item. As we will see in the following case study, clothing and personal effects can be very important for establishing identity, particularly in the absence of medical or dental records.

■**FIGURE 2:** *The outside of the temporary morgue in Orahovac, Kosovo, in 2001.*

The anthropologists start out with the basics:

- Are the remains human?

- Do they represent a single individual or several?

- What was the decedent's sex, race, stature, body weight, handedness and physique?

- Are there any skeletal traits or anomalies that could serve to positively identify the decedent?

- What is the time since death?

Unfortunately, war crimes and ethnic cleansing visit many areas that are rural, poor and politically unstable. As seen in Chapter 9, identification relies on accurate antemortem records of a putative decedent, including medical and dental records. In many areas of the world, these records do not exist, nor do we always have a clear idea about who is missing. The scattering of survivors in these war-torn areas also complicates efforts to locate families of victims – a crucial step in any effort to identify decedents using DNA.

After the biological profile is determined and preliminary efforts at identification are underway, attention is directed towards interpreting traumatic injury and determining cause and manner of death.

Documentation of Trauma

An important task facing scientists involved in the documentation of human-rights atrocities is interpreting traumatic lesions and determining whether those lesions provide evidence for homicide. Understanding how the patterns of injury differ between the victims of genocide and their enemy combatants plays a crucial role in forensic recognition of war crimes (Coupland and Meddings, 2005).

Several factors, including the number of gunshot wounds per victim, the number of projectiles recovered from each victim, the location of injuries, the trajectory or number of different trajectories, and the ratio of wounded to dead are crucial in understanding the manner of death (Warren, 2007) **(Figure 3)**. Correct interpretation helps in reconstructing the final events leading up to the death and mass burial of victims, answering such questions as:

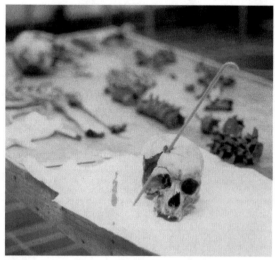

■FIGURE 3: *Investigators record the trajectory of gunshot wounds in an effort to distinguish homicide from enemy combatants killed in action.*

Bare Bones: A Survey of Forensic Anthropology

- Were the victims enemy combatants or civilians?

- Were the victims killed as a group and then transported to the gravesite?

- Is the mass grave a repository for the bodies of victims killed one or two at a time?

- Were the victims brought to the gravesite alive and then executed at the site?

The correct interpretation will provide the tribunal and family groups with information that will allow them to pursue justice.

Case Study: The Balkans

The last decade of the 20th century was difficult for people of the former Yugoslavia. The Serbo-Croat War in 1991, the Bosnian War of 1992-1995, and the subsequent advance of Serbian forces into Kosovo in the fall of 1998, pitted various ethnic groups against each other, each with – from their own perspective -- with historical claims to various areas of the former Yugoslavia (Paris, 2002). The bulk of the conflict pitted Serbian nationalists - with a highly trained and mobile remnant of the Yugoslav National Army - against Muslim Bosniaks, ethnic Albanians, and other groups in an effort to removed them from areas they perceived as their rightful homeland (Cushman, 2004). While atrocities occurred on both sides, much of the forensic work in Bosnia involved the victims of the fall of Srebrenica, where up to 8,000 Muslims were massacred over a period of nine days by Bosnian Serb forces. This event marks the largest mass murder in Europe since WWII.

Initially, forensic workers were tasked with collecting evidence for war crimes by the International Criminal Tribunal for the former Yugoslavia (ICTY), and not with identification of the victims. After the Bosnian War, starting in early spring of 1996, forensic scientists from around the world, exhumed bodies from mass graves and performed examinations of the remains in order to garner evidence for genocide throughout the Balkans.

Prior to these multinational teams working in the Balkans, forensic anthropology was unknown in the region. One consequence of these human-rights missions will be the expansion the discipline into new areas of the world, as local experts are trained in the methods employed by forensic anthropologists and archaeologists.

APPLIED RESEARCH IN HUMAN RIGHTS MISSIONS

Further research is required to obtain normative data and to establish standards for work on populations for which we do not currently have data (*e.g.*, Ross and Konigsberg, 2002. These standards should follow the ethical and moral guidelines provided by the Institutional Review Boards (IRB)s of our leading universities. In most cases, medical researchers are permitted to utilize data collected which are incidental to diagnosis and treatment in clinical practice. Similarly, forensic research can also utilize existing data, provided the identity of the decedent is not revealed. These protocols are deemed "exempt status," provided the research design is non-invasive and does not involve procedures that fall beyond those used during the normal course of a forensic investigation.

Forensic work in human-rights investigations is a relatively new discipline, so many questions remain as the scientific community develops standards and protocols applicable to international fieldwork. In addition, the teams are almost always multinational. Since the field of forensic anthropology is new to many countries, we are still trying to build relationships with pathologists and scientists in related disciplines. Among the professional and ethical questions are:

- Should there be minimum international standards for those doing the work?

- Should there be minimum international standards for the work that is done?

- Is it ethical to conduct research on remains from human-rights cases? If so, under what circumstances?

- Should we train local human-rights groups to conduct their own forensic investigations?

- What if they have no anthropological or medical background?

- To whom should remains be repatriated?

- Who owns the remains – the families or the government?

- What can/should be done when those doing the forensic investigations, or requesting the forensic investigations, are the ones who did the killing?

- What is the future of forensic anthropology and human rights (Warren *et al.*, 2004)?

Travel into war-torn areas of the world introduces new challenges to forensic workers and may entail a small amount of risk. It is imperative that visiting scientists be well-prepared by knowing something about the nature of the conflict and the culture of the people involved so they can perform their duties in a professional and diplomatic way. Good preparation reduces risks to a minimum. Many professors are working to develop courses to better prepare advanced students for human rights missions abroad (Lusiardo *et al.*, 2004). In general, however, risks are miniscule compared with the experience of learning about a new culture and contributing, even in some small way, to justice and reconciliation.

References

Cushman T (2004) Anthropology and genocide in the Balkans: An analysis of conceptual practices of power. *Anthropological Theory* 4(1):5-28.

Coupland RM, Meddings DR (1999) Mortality associated with use of weapons in armed conflicts, wartime atrocities, and civilian mass shootings: literature review. *British Medical Journal* 319:407-410.

Djuriæ M, Kjoniæ D, Rakoæeviæ Z, Nikoliæ S (2003) Evaluation of Suchey-Brooks methods for aging skeletons in the Balkans. *Forensic Sciences International* Suppl. 136:172.

Joyce C, Stover E (1991) *Witnesses from the Grave: The Stories Bones Tell*. Little, Brown and Company, 352 pages.

Lusiardo A, Drawdy S, Warren MW (2004) Desarrollo de la antropología forense y los derechos humanos en un programa universitario. *Sometido al la segunda reunión anual del Asociación Latinoamericana de Antropología Forense*.

Paris R (2002) Kosovo and the metaphor war. *Political Science Quarterly* 117(3):423-450.

Ross A, Konigsberg L (2002) New formulae for estimating stature in the Balkans. *Journal of Forensic Sciences* 47(1):165-167.

United Nations Universal Declaration of Human Rights (1948) *General Assembly resolution 217 A, III*.

UN Manual on the effective prevention and investigation of extra-legal, arbitrary and summary executions - ST/CSDHA/12 - 1991 - V. Model protocol for disinterment and analysis of skeletal remains, 1991.

Warren MW (2007) Interpreting gunshot wounds in the Balkans: Evidence for genocide. Brickley M and Ferllini R (eds.): *Forensic Anthropology: Case Studies From Europe*. Charles C. Thomas. Chapter 10:151-164.

FURTHER READING

Joyce C, Stover E (1991) *Witnesses from the Grave: The Stories Bones Tell*. Little, Brown and Company, 352 pages.

Koff C (2004) *The Bone Woman: A Forensic Anthropologist's Search for Truth in the Mass Graves of Rwanda, Bosnia, Croatia, and Kosovo*. Random House, 304 pages.

International Criminal Tribunal for the former Yugoslavia (ICTY), http://www.un.org/icty

UNMIK – Office on Missing Persons and Forensics, http://www.unmikonline.org/justice/ompf.htm

International Forensic Centre of Excellence for the investigation of Genocide (INFORCE), http://www.inforce.org.uk/index.htm

Physicians for Human Rights (PHR), http://www.phrusa.org

International Committee of the Red Cross (ICRC), "The Missing" Project, http://www.icrc.org/web/eng/siteeng0.nsf/html/themissing

UNHCHR documents on forensic science, http://www.unhchr.ch/html/menu2/i2civfsc.htm

Bare Bones: A Survey of Forensic Anthropology

Sample Test Questions

1. **The term "ethnic cleansing" as used in human rights refers to:**

   a. The systematic removal, by death or displacement, of one group of people by another group based on ethnicity, religion, or appearance

   b. Gene flow, producing admixture that makes determination of race difficult

   c. Acculturation of one ethnic group into another

   d. Ancient migration patterns that ultimately divided humans into races

2. ***Physicans for Human Rights* and the *Centre for International Forensic Assistance* are examples of:**

   a. United Nations subcommittees

   b. Non-governmental organizations (NGO)

   c. Forensic anthropology human rights teams

   d. United States governmental organizations (USGO)

3. **We know that the bodies in the mass graves of Bosnia were victims of ethnic cleansing because:**

   a. Many were found with blindfolds and ligatures still in place

   b. The large number of gunshot wounds per individual far exceeds the number expected in combatants

   c. There are thousands of dead victims, but very few injured survivors, contrary to military casualties

   d. All of the above

4. **A forensic anthropologist can contribute to Human Rights investigations in which of the following ways?**

    a. Excavation

    b. Identification

    c. Trauma analysis

    d. All of the above

16

The Profession of Forensic Anthropology

How does one become a forensic anthropologist? Practitioners of forensic anthropology require an advanced degree in biological anthropology. A few programs that provide training in forensic identification exist outside of anthropology (*e.g.*, human biology), but as a practical matter these programs are closely related to biological anthropology programs, in that they place emphasis on skeletal biology and are under the guidance of board-certified forensic anthropologists (most notably the University of Indianapolis).

A minimum of a Masters degree is required to practice forensic anthropology. It is doubtful that a basic level of competency could be achieved with less formal training. An advanced degree also serves as a qualification or credential, insomuch as practicing forensic anthropologists must testify in court and opposing legal counsel quickly find a more qualified expert to impugn a less qualified expert's testimony.

Most forensic anthropologists hold a Ph.D. degree – the highest academic degree in America. The *American Anthropological Association* (AAA) reports that the average time for completion of the Ph.D. after the baccalaureate is 8.5 years, with the average age of the recipient being 39 years old (AAA website, 2007). Of the approximately 400 Ph.D.s in anthropology awarded annually, around 10% (40) are in biological anthropology, and perhaps 10% of those recipients have research interests in forensic identification and trauma analysis.

How can undergraduates interested in becoming forensic anthropologists realize their dream? They can start by getting a baccalaureate degree in anthropology at a good undergraduate institution. It should be noted that a degree in biology, chemistry or related field will not disqualify applicants from consideration. This is one unique aspect of anthropology graduate studies! Since anthropology often operates at the margins of other disciplines, most graduate programs are open to the possibility of accepting students from outside the discipline, as long as their prior studies relate to an area of research interest within anthropology. Grades and test scores must be competitive, because interest in forensic anthropology is high and the best graduate programs receive large numbers of applications. Try to seek broad training at the Masters level and attempt to get at least some experience in and around a working forensic anthropology laboratory.

OPPORTUNITIES IN FORENSIC ANTHROPOLOGY

As described in the preceding paragraph, few doctorates are awarded to forensic anthropologists every year. With the Ph.D. in hand, several opportunities are available for employment.

- Academia

- Medical Examiner's Offices

- Laboratories (JPAC – also a governmental organization)

- Non-governmental organizations

- Governmental organizations (NTSB; AFIP; DMORT)

- Museums

- Private consulting

Academia

Forensic anthropologists have traditionally sought academic positions in universities and colleges. Academic institutions permit one to pursue research interests within the field and provide the laboratory space, equipment and supplies needed to recover, process and examine human skeletal material. College professors work within the boundaries of the three missions of higher education: teaching, research and service. University-based forensic anthropologists usually consult with medical examiners and coroners as part of the service component of their positions, much like other professors might serve on the faculty senate, do agricultural outreach, or serve on committees that develop public policy.

Medical Examiner's Offices

Chief Medical Examiners are increasingly realizing the contribution that forensic anthropologists can make to their caseload, not only by performing skeletal analyses, but also by functioning in multiple roles as medicolegal investigators, autopsy assistants, or trace analysts. Some of the larger offices employ colleagues solely in the capacity as forensic anthropologists. These offices have the resources and large caseloads that make hiring a full-time forensic anthropologist a logical decision. Medical Examiners are finding that anthropologists are valuable in helping assess traumatic injuries to bone in more routine autopsy cases **(Figure 1)**.

■FIGURE 1: *Dr. Laura Fulginiti of the Maricopa County, Arizona Office of the Medical Examiner, shows investigators a human bone excavated at a crime scene.*

Laboratories (JPAC)

The United States government allocates considerable resources to locating, recovering and repatriating members of the military that were killed in action during past military conflicts. This effort is accomplished by the Joint POW/MIA Accounting Command, or JPAC. The *JPAC's* Central Identification Laboratory in Hawaii remains the single largest employer of forensic anthropologists in the world. The Central Identification Laboratory also offers opportunities for pre-doctoral and post-doctoral Fellowships through the Oak Ridge Institute for Science and Education (*ORISE*). These *ORISE* Fellows work side by side with JPAC anthropologists in the field and laboratory in preparation for future employment, either at the JPAC laboratory or in academia or Medical Examiner's offices (**Figure 2**).

■**FIGURE 2:** CIL *Anthropologist Paul Emanovsky working in the Xepon River Valley, Laos. Photograph by Cpl. Nicholas Riddle.*

Non-governmental organizations (PHR, CIFA INFORCE, ICRC, etc.)

Non-governmental organizations, or NGOs, are groups dedicated to serving humankind and include such diverse entities as the Red Cross, the Salvation Army, or Amnesty International. As we described in Chapter 16 - *Human Rights Missions*, several NGOs have, through circuitous routes, become involved in the investigation of human rights abuses, genocide and war crimes. These organizations include *Physicians for Human Rights (PHR)*, the *Centre for International Forensic Assistance (CIFA)*, and the *International Forensic Centre of Excellence (Inforce)*.

Governmental organizations

As noted in Chapter 15, some forensic anthropologists choose to become involved in the recovery and identification of victims of mass disasters, during which time they become federal employees during their deployment. However, anthropologists are also employed in other governmental agencies as well.

One example is the National Transportation Safety Board (*NTSB*), located in Washington, D.C. The NTSB is an advisory agency that seeks to make commercial travel safer and is responsible for investigating and determining the cause of transportation disasters. Among the various support services, the NTSB provides staff with expertise in victim identification in order to coordinate identification efforts among various responding agencies tasked with victim identification, including *DMORT*. The safety board employs a full-time forensic anthropologist, Paul Sledzik, as manager of victim recovery and identification within the Office of Transportation Disaster Assistance (*TDA*). The *TDA* provides assistance to survivors, families and communities affected by an air traffic disaster. Mr. Sledzik has extensive experience working to identify victims of mass disaster and is a former *DMORT* commander.

Museums

Some forensic anthropologists are employed as curators or collection managers at museums. The Smithsonian Institution provides forensic services to the Federal Bureau of Investigation and other governmental agencies, as well as maintaining the skeletal collections that have provided normative data for many of the methods used in forensic identification. Anthropologists at the Armed Forces Institute of Pathology's National Museum of Health and Medicine also maintain collections that are valuable for research in trauma and skeletal pathology, and also conduct a well-known short course in forensic anthropology co-sponsored by the American Registry of Pathology.

Private Consulting and For-Profit Organizations

At this time there does not seem to be enough demand for consultation and expert testimony, outside of sources in academia and within the medical examiner's office, to make private consulting a full-time career path. Many forensic anthropologists, especially those within the university system, are just beginning to charge fees for their services. Therefore, it is difficult for private consultants to find cases in which other anthropologists would not consult without charge. Even when academic-based anthropologists are provided a fee for their services, many opt to place those funds into foundation or research accounts to help fund their student's research activities or laboratory expenses.

Still, some colleagues successfully consult on a fee-for-service basis regularly, and several are active in civil cases – many involving disputed cremains (see chapter 14).

In recent years, a few for-profit corporations have become involved in forensic investigations and disaster response, providing diverse services on a contract basis with local, state and federal governments. Among those is Kenyon International, a corporation that provided disaster management services following Hurricane Katrina, the 2005 tsunami, and the World Trade Center attacks. Forensic anthropologists are deployed by Kenyon in much the same way as the DMORT teams.

PROFESSIONAL ORGANIZATIONS

Most academic forensic anthropologists belong to the *American Anthropological Association (AAA)* and the *American Association of Physical Anthropologists (AAPA)*. The *AAPA*, in particular, provides the forum for most new research in human osteology, skeletal biology, bioarchaeology and related interests shared by most forensic anthropologists. The journal for the *AAPA*, the *American Journal of Physical Anthropology*, is the most prestigious journal within the subfield and forensic practitioners will find that many advances in skeletal analysis are published here.

The *American Academy of Forensic Sciences (AAFS, or "the Academy")* is home to the top forensic scientists in the country, including forensic anthropologists. The *AAFS*, founded in 1948, has over 6,000 members representing every state and province in the United States and Canada, as well as over 55 other countries. The academy is ". . . dedicated to the application of science to the law;" and accomplishes its goal by disseminating information and research through meetings, newsletters, and its internationally recognized *Journal of Forensic Sciences* (*AAFS* website, 2007). The annual meetings feature over 500 papers, workshops and special programs and provide a place where scientists can gather and exchange ideas and keep up to date on the most current information and research. Every annual meeting also features a "student academy," in which local high school and undergraduate students can attend and learn about the forensic sciences.

Members of the *AAFS* are grouped by areas of specialization into 10 sections: Criminalistics, Engineering Sciences, General, Jurisprudence, Odontology, Pathology/Biology, Physical Anthropology, Psychiatry/Behavioral Sciences, Questioned Documents, and Toxicology. Each section has various levels of membership, beginning with Student Affiliate, Trainee Affiliate, Associate Member, Member, and finally, Fellow.

Forensic anthropologists are, naturally, in the Physical Anthropology section of the academy. This section was founded in 1973 by several like-minded anthropologists who sought formal recognition for the discipline and wished to establish professional and ethical standards for future generations of practicing forensic anthropologists. As mentioned in Chapter 1, the establishment of the Physical Anthropology section marks the beginning of the modern era of our discipline. This event, along with the new certification process outlined below, firmly rooted forensic anthropology as an established member of the forensic sciences.

There are two other organizations dedicated to improving the forensic sciences: The *International Association for Identification (IAI)* and the *International Association of Forensic Sciences (IAFS)*. The *IAI* was formed in 1915 and is the oldest and largest forensic science/forensic identification organization in the world. This organization is over 6,700 members strong. The organization publishes the *Journal of Forensic Identification* bimonthly. Many members of the *AAFS* are also members of the *IAI*. Inaugurated in 1957, the *IAFS* is the only *worldwide* association that brings together academics and practitioners of various forensic disciplines. The IAFS holds triennial meetings with the assistance of the AAFS staff. Just this year, the organization added a forensic anthropology division.

CERTIFICATION AS A FORENSIC ANTHROPOLOGIST

Most practicing forensic anthropologists ultimately seek accreditation by the American Board of Forensic Anthropology (*ABFA*). The *ABFA* is a non-profit organization recognized by the American Academy of forensic Sciences as the foremost accrediting agency for the forensic anthropology profession. The *ABFA* was established in 1977 when the founding members of the Physical Anthropology Section of the *AAFS* met and established the criteria to be used for certification. The objectives of the Board, as cited on the *ABFA* website, are:

(a) to encourage the study of, improve the practice of, establish and enhance standards for, and advance the science of forensic anthropology; (b) to encourage and promote adherence to high standards of ethics, conduct, and professional practice in forensic anthropology; (c) to grant and issue certificates, and/or other recognition, in cognizance of special qualification in forensic anthropology to voluntary applicants who conform to the standards established by the Board and who have established their fitness and competence thereof; (d) to inform the appropriate branches of federal and state governments and private agencies of the existence and nature of the *ABFA* and the professional quality of its Diplomates for the practice of forensic anthropology; (e) to maintain and furnish lists of individuals who have been granted certificates by the Board. In this way the *ABFA* aims to make available to the judicial system, and others, a practical and equitable system for readily identifying those persons professing to be specialists in forensic anthropology who possess the requisite qualifications and competence (*ABFA* website, 2007).

Eligibility for certification by the *ABFA* is based on the candidate's education, training and experience, as well as on results of formal practical and oral examinations administered during the annual meeting of the *AAFS*. The *ABFA* currently requires that a candidate be a permanent resident of the United States, Canada, or their territories; and must hold an earned doctoral degree in Anthropology with an emphasis in Physical Anthropology. Prior to sitting for the board examination, the applicant must have at least three years of full-time professional experience and must submit examples of case reports. All applicants must also document a record of contribution to the field of forensic anthropology in the form of published research or scientific presentations at professional meetings.

Once all criteria have been met, the successful candidate becomes a *diplomate* of the *ABFA*. At the time of publication of this volume, there are 75 diplomates of the *ABFA*, among whom about 65 are still active. Every board-certified forensic anthropologist in North America is a Member or Fellow of the *AAFS*.

THE FUTURE OF FORENSIC ANTHROPOLOGY

In 2007, at the 59th annual meeting of the American Academy of Forensic Sciences in San Antonio, Texas, a symposium was organized within the Physical Anthropology section titled *The Fourth Era of Forensic Anthropology* (Sledzik *et al.*, 2007).

During this session, the Fourth Era of forensic anthropology (with the year 2000 marked as the beginning) was described as an *expansion phase* manifested by an upsurge in student interest and media attention, an increase in the number and diversity of jobs related to the discipline, a dramatic escalation in the number of universities and colleges offering undergraduate courses in forensic anthropology, and the development of several new graduate programs offering a focus in forensic anthropology.

The first graduate programs to design specialized curricula to specifically train forensic anthropologists appeared in the 1970s at the University of Arizona, the University of Florida, the University of New Mexico, the University of South Carolina, and the University of Tennessee. Described by Sledzik as the *establishment phase*, these programs trained many of the current cohort of practicing forensic anthropologists, some of whom have gone on to establish newer programs at universities such as California State University at Chico, the University of Indianapolis, Mercyhurst College, and several others

(Sledzik *et al.*, 2007; *for a list of universities and colleges with programs in forensic anthropology, see Appendix 1*).

The ever-increasing roles of forensic anthropologists, including increased collaboration in assessing skeletal trauma in fleshed cases, estimating post-mortem interval, evaluating taphonomic modification of bone, and entry into the investigation of war crimes and mass disasters, are requiring our educational programs to reassess the graduate curriculum and to provide more specialized coursework. Our graduate students now make forays into medical and dental schools, law schools, and other non-traditional venues in order to obtain the knowledge base necessary for today's forensic practice. At the same time, we are obliged to continue to train anthropologists in traditional physical anthropology programs and encourage broad-based dissertation topics that will lead to employment in academia, so we will have mentors in place to train future generations of forensic anthropologists.

The future of forensic anthropology will also present ethical issues as we strive to protect the rights of decedents and their families, but also train and educate our students and conduct research to solve evolving questions. Current pressing ethical questions include:

- Is it permissible use active forensics cases as a teaching tool?

  Most graduate programs in forensic anthropology follow a medical-school model of "grand rounds," which involve students directly in casework. Advanced graduate students interested in human identification are pre-professional students in a sense that they are training for forensic anthropology practice. A critical component of their training is to insure that they are knowledgeable about guidelines that maintain confidentiality. So, as with students of clinical medicine, it is possible to protect the rights of the decedent, yet train the next generation of forensic anthropologists.

- Does institutional review designed to protect patient's or research subject's rights suffice to protect the rights of the decedent?

  Most research proposals in skeletal biology are considered exempt status protocols by institutional review boards (IRBs), which normally deal with clinical trials involving living subjects. The IRBs provide strict guidelines for how results are to be reported. These guidelines insure that the identity and dignity of the decedents is protected.

- What circumstances exist that would prohibit the use of active or resolved cases as a resource for the classroom?

Since NAGPRA legislation has decreased the volume of available skeletal material for teaching and research, photographic images of skeletal material have become a more important resource for teaching skeletal anatomy – a basic foundation for further study in skeletal biology, forensic identification and related disciplines. In 2001, Dale Earnhardt, a professional race car driver, was killed on the last lap of the Daytona 500 auto race. Since the death was "unattended," the case fell under the jurisdiction of the local medical examiner. Florida's liberal "Sunshine" statutes ensure public access to government records including autopsy protocols and photographs Amid family concerns about privacy, the Florida legislature quickly enacted a law excluding autopsy photographs, videos and audio recordings from the requirements of Florida's disclosure statures. However, the new legislation does not provide a clear definition of what constitutes an "autopsy photograph" or what is meant by the "public." Are pathologists able to present case studies at professional meetings? Are they able to publish photographs of cases in scientific journals? Are the photographs taken to document skeletal cases by anthropology consultants considered autopsy photographs? If so, an unintended consequence of this new legislation would be to restrict the use of photographic images of cases, which have become an important teaching adjunct, especially for undergraduate students who are not permitted access to actual casework (Hotzman and Warren, 2002).

As we have shown in several of the preceding chapters, much of our landmark research has been conducted using anatomical specimens harvested during autopsy. Most state laws governing medical examiners permit the retention of specimens for further analysis if that analysis is beneficial regarding public health issues or the case at hand. In the past, we have not obtained consent from the family before obtaining specimens for research, since forensic research is deemed to be vital to public health and safety – an important consideration within the realm of the Medical Examiner's and Coroner systems. The public is becoming better informed about the need for autopsy and medical investigation, including television shows that detail autopsy procedure and death investigations (*e.g.*, *Ask Dr. Baden, Skeleton Stories, and Dr. G - Medical Examiner*). How can we expand our knowledge of contemporary populations without

appropriate skeletal samples? How can we obtain consent when we don't know the identity of the decedent? Anatomical specimens are often kept as biological evidence of identity, or retained as exhibits of trauma for courtroom testimony. In cases where the decedent is identified and returned to the family, what should be done with these specimens? Should these specimens be available to students and faculty for instruction or research?

There are no clear-cut correct answers to these questions. Medical ethicists, institutional-review boards and the public will grapple with many of these issues for years to come. In the past, we seem to have always reached a balance between individual rights and the interests of the public at large.

References

American Anthropological Association website (2007)
http://www.aaanet.org/surveys/97survey.htm

American Board of Forensic Anthropology website (2007)
http://www.csuchico edu/anth/*ABFA*/#Background

International Association for Identification website (2007). http://www.theiai.org/

Sledzik PS, Fenton TW, Warren MW, Byrd JE, Crowder C, Drawdy SM, Dirkmaat DC, Finnegan M, Fulginiti LC, Galloway A, Hartnett K, Holland TD, Marks MK, Ousley SD, Rogers T, Sauer NJ, Symes SA, Tidball-Binz M, Ubelaker D (2007) The Fourth Era of Forensic Anthropology. *Proceedings of the American Academy of Forensic Sciences* 13: 350–353.

Hotzman JL, Warren MW (2002) Earnhardt Crashes into Florida Sunshine: Consequences of SB 1356 legislation on the teaching of forensic anthropology. A paper presented at the 101st annual meeting of the American Anthropological Association, New Orleans, Louisiana. *Abstracts of the American Anthropological Association*, 2002.

Sample Test Questions

1. **Professional forensic anthropologists must have:**

   a. Taken a course in forensic anthropology

   b. A minimum of a Bachelor's degree

   c. A minimum of a state-sponsored short course in forensic identification

   d. A minimum of a Masters degree, and usually a Doctorate

2. **Professional forensic anthropologists work in:**

   a. A university or college

   b. A medical examiner's office

   c. A laboratory like the Central Identification Laboratory, JPAC

   d. All of the above

Appendix

There is currently no accreditation process to evaluate and standardize curricula in forensic anthropology. Therefore, educational programs are inconsistent in terms of curriculum and faculty qualifications. The following schools were submitted to the Physical Anthropology section of the American Academy of Forensic Sciences by an *adhoc* committee on education, consisting of Dr. Ann Ross (Chair), Dr. John Byrd, Dr. Tom Crist, and Dr. Vincent Stefan.

Graduate Programs in Anthropology with section members on faculty and specified Forensic Anthropology Masters and/or doctoral concentrations:

California State University, Chico
School of Graduate, International &
Interdisciplinary Studies
400 West 1st
Chico, CA 95929-0875
530-898-6880

Mercyhurst College
Graduate Program in Forensic and
Biological Anthropology
501 East 38th Street
Erie, PA 16546
814-824-2000

Michigan State University
The Graduate School
110 Linton Hall
East Lansing, Michigan 48824-1044
517-353-3220

University of Florida
Office of Admissions
201 Criser Hall
PO Box 114000
Gainesville, FL 32611-4000
352-392-1365

University of Hawaii at Manoa
Graduate Division
2540 Maile Way
Spalding Hall 360
Honolulu, HI 96822
808-956-7541

University of Indianapolis
Graduate Program in Human Biology
University of Indianapolis
1400 E. Hanna Ave.
Indianapolis, IN 46227-3697
317-788-3486

University of Tennessee, Knoxville
Graduate and International Admissions
201 Student Services Building
Knoxville, TN 37996-0230
865-974-3251

University of Central Lancashire
Admissions Office
Preston PR1 2HE
United Kingdom
+44-0-1772-201201

Graduate Programs in Anthropology w/ section members on faculty who could "Potentially" mentor a Forensic Anthropology thesis/dissertation:

Arizona State University
Division of Graduate Studies
Administration B-Wing, Room 285
PO Box 871003
Tempe, AZ 85287-1003
480-965-6113

Binghamton University, State University of New York
Graduate School
PO Box 6000
Binghamton, New York 13902-6000
607-777-2151

North Carolina State University
The Graduate School
1575 Varsity Drive, Flex Lab, Module 6
Campus Box 7102
Raleigh, NC 27695-7102
919-515-2872

George Washington University
Graduate Program in Anthropology
2110 G Street
NW Washington, DC 20052
202-994-6075

The Ohio State University
Graduate School
250 and 247 University Hall
230 N. Oval Mall
Columbus, OH 43210-1366
614-292-6031

Simon Fraser University
Department of Sociology and Anthropology
8888 University Drive
Burnaby, British Columbia
Canada V5A 1S6
604-291-3518

Bare Bones: A Survey of Forensic Anthropology

Texas State University, San Marcos
The Graduate College
601 University Drive
San Marcos, TX 78666-4605
512-245-2581

Texas Tech University
Graduate School
02 Holden Hall
PO Box 41030
Lubbock, TX 79409-1030
806-742-2781

Tulane University
Graduate Program in Anthropology
1021 Audubon Street
New Orleans, Louisiana 70118
504-865-5336

University of Alabama at Birmingham
UAB Graduate School
Hill University Center 511
1530 3rd Avenue South
Birmingham, AL 35294-1150
205-934-8227

University of Calgary
Faculty of Graduate Studies
Earth Sciences, Room 720
2500 University Drive NW
Calgary, Alberta, Canada T2N 1N4
403-220-5417

University of Central Florida
Division of Graduate Studies
Millican Hall, Suite 230
PO Box 160112
Orlando, FL 32816-0112
407-823-2766

University of Missouri, Columbia
MU Graduate School
210 Jesse Hall
Columbia, MO 65211
573-882-6311

University of Nevada, Las Vegas
Graduate College
FDH 352, Box 451017
4505 S. Maryland Parkway
Las Vegas, NV 89154-1017
702-895-3320

University of New Mexico
Office of Graduate Studies
MSC 03 2180
1 University of New Mexico
Albuquerque, NM 87131
505-277-2711

University of North Texas
Toulouse School of Graduate Studies
P.O. Box 305459
Denton, TX 76203-5459
888-868-4723

University of Southern Mississippi
Office of Graduate Studies
118 College Drive #10066
Hattiesburg, MS 39406-0001
601-266-4369

University of Texas at Austin
Office of Admissions / GIAC
2608 Whitis Avenue
P.O. Box 7608
Austin, TX 78713-7608
512-475-7390

University of Toronto
School of Graduate Studies
65/63 St George Street
Toronto, Ontario
Canada M5S 2Z9
416-978-6614

University of Wyoming
UW Graduate School
Knight Hall, Room 109
1000 E. University Avenue, Department 3108
Laramie, WY 82071-3108
307-766-2287

Wichita State University
Graduate School
1845 Fairmount
Wichita KS 67260-0004
316-978-3095

Graduate Programs in Forensic Sciences:

Florida International University
Department of Chemistry
Miami, FL 33199
305-348-6211

Marshall University
Forensic Science Center
1401 Forensic Science Drive
Huntington, WV 25701
304-690-4363

Virginia Commonwealth University
College of Humanities & Sciences
1000 West Franklin Street
PO Box 843079
Richmond, VA 23284-3079
804-828-8420

Michigan State University
560A Baker Hall
East Lansing, MI 48824-1118
517-353-7133

University of Alabama at Birmingham
Department of Justice Sciences
UBOB 210
1201 University Blvd.
205-934-2069

George Washington University
Samson Hall 101
Department of Forensic Science
2036 H Street, NW
Washington, DC 20052
202-994-7319

John Jay College of Criminal Justice
445 West 59th Street
New York, NY 10019
212-237-8899

INDEX

A

advancing age 161–162
African ancestry 118, 153
Age of Discovery, The 19
Algor mortis 50
American Academy of Forensic Sciences (AAFS) 24
American Association for the Advancement of Science (AAAS) 25, 270
American Board of Forensic Anthropology (ABFA) 24, 288–289
American Society of Crime Laboratory Directors-Laboratory Accreditation Board 26
Amerindian 120–121
anatomical structures 142, 163, 172, 176, 201, 211
anatomical variants 172
anatomists 21, 24, 66, 69
ancestral affiliation 114–115, 186
ancestry 23, 60, 81, 83, 94, 101, 111–122, 151–153, 172, 260
animal
    predation 51–53
    carnivore 53–55
    rodent 53–55
Antemortem 175–178
Anterior 66–67, 70
anterior femoral curvature 121
anthropologist 127
anthropology 15–23
    archaeology 16, 40
    biological 281–292
    formative period 20–21
    linguistic 16
    socio-cultural 16
    theory 15, 17

arm
    humerus 78, 81, 86
    proximal joint 78
axial skeleton 68
axis 67, 70, 217

B

Bioarchaeologists 24, 141
biohazardous material 45
biological
    concept of race 111
    determinism 113
    markers 169
    profile 169
    sex 93–107
bone samples 11
bones 18, 32, 35, 53, 56, 58, 80–82
Bregma 84, 101, 116–119, 122

C

cadaver
    dogs 33, 40
    stature 155
calcined 57, 59, 236, 239–241, 255
cameras 193
case logbook 43
Central Identification Laboratory 23, 284
cervical vertebrae 69, 72
chemical analyses 44, 245
children, prepubescent 94
clandestine burials 37–40
Cleveland Museum of Natural History 21
computed axial tomography 66
Computerized Tomography (CT) 176
consolidation period 20, 22

coronal plane 66
coroner system 6
corpus delicti 170
costal cartilage 74, 100
cranial
    features 176
    measurements 119
crest supramastoid 98
Creutzfeldt-Jakob Disease 10
crown-heel length (CHL) 130
crushing fractures 53

D

damage 22, 38, 56
    fire 56, 58–60
    intense heat 58–60, 261
decomposition 24, 40, 44
decomposition natural 4
degeneration 139
delamination of bone 57
dental disease
    caries 140
    periodontitis 140
    wear 141
dentition 132, 139, 260
    deciduous teeth 132
    permanent teeth 132
deoxyribonucleic acid 172
determination of
    ancestry 114
    sex 94, 171, 186
        metric assessment 95, 100
        nonmetric assessment 95
development 26, 65, 101, 128, 142
diaphyseal length 129–131
diaphysis 77–78, 177
digitalization 176
discrete
    races 114
    units (now known as genes) 112
DNA testing 10

Dogs and coyotes 53
dorsal pitting 104

E

East asians 118
electronic video mixing board 193
Ellis Kerley Foundation, The 26
entomologists 11, 50
environmental process 56
    weathering 57
essentials of Forensic Anthropology 24
estimated stature 164
Eugenics movement, The 20, 113
European ancestry 118
Evidence 42–45
examination of skeletal and burned
    remains 21

F

factual information 9
FBI Law Enforcement Bulletin report 28
femoral head 100, 103–104, 137–138
femur 81, 86, 121
fetal
    growth 128–129
    length 128–129
    period 127–128, 146
Fordisc discriminant function computer
    program 85, 102
forearm
    radius 67, 78–79, 131, 153, 171
    ulna 52, 79, 213–214,
forensic
    anthropologists 11, 23–24, 40, 65,
        101, 114, 281–290
    anthropology 15–22, 269–270, 275,
        277, 288
    art 193–200
    Data Bank 24–25, 101, 120, 154
    Education Program Accreditation
        Commission (FEPAC) 26

Bare Bones: A Survey of Forensic Anthropology

odontologist 11, 240, 256
    stature 153
Fourth Era, The (2000-present)
    26, 289
fragmentary long bones 163
frontal sinuses 177

G

gestational age of a fetus 130
global positioning system (GPS) 41
ground-penetrating radar 38–39
growth 65, 77, 127–129

H

H. afer 112
H. americanus 112
H. asiaticus 112
H. europus 112
Haase's rule 129
Hallux 82
Hamann–Todd collection 21,
    100, 154
health hazard 4
histological examination 10
Holmes Oliver Wendell 21
home funeral 4
human
    anatomy 193
    populations 152
    Scatter
        intentional 56
        unintentional 55
    skeletons 22
Human Skeleton in Forensic Medicine,
    The 23

I

identity of the decedent 6, 8, 46, 127,
    170, 193, 276, 292
iliac crest 138

in situ 36, 271
infrared photography 40
inion hook 98
investigators
    crime scene 11
    death 7–8, 11
    homicide 8, 170
    medicolegal 3, 56, 282

J

joints
    cartilaginous 52
    fibrous 52
    synovial 52

K

Kerley Award, The 26

L

lateral 67, 73–76, 79, 82, 96, 226
left eye orbit 66
leg
    ball and socket joint 81
    Femur 81, 100, 104, 121
    Fibula 67, 82, 153
    patella 82
    Tibia 82, 121, 153
life-history 171
life history markers 173
life-history variables 172–173
linea aspera 100
linear-regression formulae 130, 163
living stature 155
long bones 53, 67, 77, 81, 128, 131,
    153, 163, 217

M

maceration 43, 45
magnetic resonance imaging
    (MRI) 176

Bare Bones: A Survey of Forensic Anthropology

search methods
  circular 34
  grid 34
  invasive 33
  line 34
  noninvasive 33
service memorial 4
sexual
  dimorphism 153
  selection 93
Shattuck Lecture 21
shoulder girdle 75
  Clavicle 75–76
  scapula 75–76
skeletal
  age 142
  biologists 24
  framework 139
  remains 21, 23, 31–34, 37, 51, 55,
    57, 114, 151
skeletonization 50
Smithsonian Institution in
  Washington, D.C. 22
soil staining 57
stature 151
sternum (or breastbone) 74
  body 74
  the manubrium 74–75
  the xiphoid process 74
stress
  environmental 94, 173
  nutritional 94, 173
sun bleaching 57
sutures
  sagittal 116, 219
  supranasal 116, 118
  transverse palatine 116–118
symphyseal face 142–143

T

taphonomy 49–60
tarsals 82
terry collection 22
thoracic vertebrae 71–72
three-dimensional facial
    approximation 197
time of death 50, 60, 213–214
time since death 50, 175, 271
toxicology 10, 287
traits
  discontinuous 115
  nonmetric 115
  quasicontinuous 115
transverse plane 66
trauma
  perimortem 214
  postmortem 215
two-dimensional art 196
typologies 20

U

unattended deaths 3

V

vertebrae 68–73
video superimposition 193
vitruvian man 151

W

warping 59, 238
water transport
  fluvial transport 58
World War II 23, 153, 269
wrist
  carpals 80, 82, 134
  metacarpals 80, 82, 134
  phalanges 80, 82–83, 134